THE FARMER'S ANIMALS

THE FARMER'S ANIMALS

How they are bred and reared

BY

FRANK H. GARNER

*M.A., M.Sc., University Lecturer
in Agriculture. Agricultural
Organizer for East Suffolk*

WITH SIXTEEN PHOTOGRAPHS

CAMBRIDGE

At the University Press

1943

CAMBRIDGE
UNIVERSITY PRESS

University Printing House, Cambridge CB2 8BS, United Kingdom

Published in the United States of America by Cambridge University Press, New York

Cambridge University Press is part of the University of Cambridge.

It furthers the University's mission by disseminating knowledge in the pursuit of
education, learning and research at the highest international levels of excellence.

www.cambridge.org
Information on this title: www.cambridge.org/9781107629509

First published 1943
First paperback edition 2014

A catalogue record for this publication is available from the British Library

ISBN 978-1-107-62950-9 Paperback

CONTENTS

NOTES ON THE PHOTOGRAPHS

*In the text page references to pictures are given
in black type, thus* (p. 81)

Note (*a*) the straw-covered hurdles round the outside to afford shelter, (*b*) cubicles against the outside hurdles into which ewes and lambs may be placed, (*c*) the yard is divided into various sections for ewes with singles or with twins, and for unlambed ewes, (*d*) troughs in the foreground, and hay racks in the background, (*e*) these lambs have been tailed.

The sheep is seated upon its haunches by the shearer and the wool shorn off by cutting first one half of the fleece from ·the belly to the backbone (as shown at the left) and then the other half, again by shearing from the belly to the backbone. Clipping is done as close as possible to the skin, the shears taking strips ½ in. to 1 in. wide with each cut.

Note (*a*) each hut and run is self-contained, (*b*) all meal and water must be provided on the site, (*c*) pigman is taking water from the galvanized tanks, (*d*) the huts and runs can be moved forward by the simple hoist (a pair of old car wheels) shown in the foreground, (*e*) these huts are moved to fresh sites at least twice a week, (*f*) with this system the pig manure is left on the field, and no cleaning of pens is required.

Up to 100 pigs can be kept in one yard provided (*a*) there are sufficient troughs for all to feed *at any one time*, (*b*) the pigs are all about the same size, (*c*) there is room for all pigs to sleep under cover at night, (*d*) all pigs remain healthy, and any sick pig is removed as soon as it is ill. These same yards with slight modifications can be used for other stock. See note to photograph 3.

Note (*a*) two rows of pens each for about ten pigs, (*b*) the central passage for feeding, (*c*) the dunging passages along the outside walls, (*d*) the many windows, (*e*) the absence of an upper storey, which is present in the piggeries in Denmark, (*f*) for the sake of cleanliness, concrete is extensively used, (*g*) this type of piggery can be modified for sows and their litters, but for the best results with litters there must be a run outside the piggery for each pen, so that the pigs can enter and leave at will.

So long as care is exercised when additional mares and foals are first put into a field it is quite safe to have a number of mares and their foals running together, or even with horses that are not breeding. Some of the mares in this picture are of the light breeds and others of the heavy breeds.

This horse is obviously well under control. Typical harness is being used, including the long reins one of which the groom is holding in each hand.

There must be many passages between brooders to make it easy to fill troughs with meal or water. Water can be seen in the metal troughs at the left of the photograph, and meal is in the larger troughs at the right. To save the labour of carrying water the troughs are filled from the hose attached to the main water supply overhead.

The poultry units must be moved to fresh grass every few days; these units are for use when for any reason (possibly because of foxes) the birds are not to be allowed outside the runs. Water, grit, and meal are kept in troughs constantly before these birds. These same units can be used for laying birds, fattening birds, or chickens over two months of age. The white-faced sheep are Cheviot × Border Leicester, the black-faced sheep are Suffolks, and those with wool on their black faces are probably crossbred. Sheep are very useful for grazing around poultry pens and houses because they do not damage the equipment.

Each pen has its own water, grit, and meal (note the three troughs in front of each pen). When eggs are laid they roll forward so that the hens cannot eat them.

While the goats are grazing the boys are collecting grass for the goats' evening meals. Goats will eat much food that would otherwise be wasted.

PREFACE

This book has been written to meet a definite and growing need. More and more young people, not only in rural areas but also in the towns and cities, are taking interest in aspects of agriculture, and this tendency has undoubtedly been increased by the war, with its inevitable emphasis upon the sources of the national food supply. It was found, however, that very few of the existing books on agriculture are really suitable for even the older of such readers; thus this book came to be designed for them. Members of Young Farmers' Clubs may also find it a handy introduction to the subject. Its main object is to awaken young people's interest in livestock and the art of managing them. Once interest has been aroused there is a wide range of more advanced books available for study.

As the book had to be short, information about the various breeds has been kept to a minimum so as to give as much space as possible to general principles applicable to many breeds. Feeding, too, could only be discussed here in elementary fashion; it is a subject so important that it has already many books entirely devoted to it. Reference has occasionally been made to certain ways in which the present war influences the keeping of livestock, but in general the book is written with an eye to permanent rather than to temporary conditions.

Special thanks are offered to the various teachers and residents in East Suffolk (far too many to mention by name) for their constructive criticism, to my wife for assistance in reading the script and proofs, to the staff of *The Farmer and Stockbreeder* for help so willingly given in finding and lending suitable photographs, and finally to the publishers for their patience and invaluable assistance.

F. H. G.

May 1943

CHAPTER I. THE USES OF
FARM ANIMALS

NOBODY taking a country walk among the lanes and fields and farms of England will be surprised to come across cows or horses, sheep or pigs, cocks and hens, or goats. Such stock are in fact so commonplace and expected that most people rather too easily take them for granted.

It may at first seem almost too simple a question to ask why animals are kept on farms. Nevertheless the whole of this first chapter will be needed to give the proper answer, which is by no means simple, or perhaps so obvious as it seems.

To begin with, we should remember that farm animals are not native creatures of the countryside at all; and a simple thought about the reason for gates and fences, and yards and fields, would prove as much. They are kept by the farmer at much cost and trouble to himself. He and his men care for these beasts with hard work in all weathers; yet if anyone should ask the farmer why he takes so much trouble he might very well reply with a laugh: 'I don't keep the animals; I expect the animals *to keep me.*' A farmer has to live and to pay his men, and certainly he makes part of his living by breeding and rearing and selling his *stock*, which is the general name for farm animals: he keeps them for the rest of us; we see them again, in fact, as food on our tables. But though the first part of this chapter will be about what animals produce, this is by no means the whole of the story, as we shall see.

WHAT ANIMALS PRODUCE

Meat. All farm animals in Britain except horses provide us with meat. In ordinary times about half of the meat eaten in Britain is produced here; the rest comes from overseas: beef principally from the Argentine; mutton and lamb from Australia, New Zealand, South Africa, and the Argentine; bacon and pork from Denmark, Holland, Canada, and some countries of Central Europe; and poultry mainly from Europe also.

English meat comes from stock kept in many parts of the country. Sheep, for instance, are to be found in all districts, though they cannot be fattened for meat where soil is poor, because sheep are generally fed on natural foods, and poor soil will not produce crops good enough.

Cattle are fattened for beef either on the good grassland in the Midlands and on the marshes found up and down the country, or in yards in arable districts, where they can be fed on crops grown near at hand, such as roots, hay, straw, and different grains.

Pigs are not so widely kept as sheep and cattle. They are often fed upon 'waste foods'—skimmed milk (left over from cream and butter making), whey (left over from cheese making), potatoes not good enough for human food, and poor grain. Poultry and pigs are found in much the same places, because for the most part they are fed on the same kinds of foods.

Milk.   In the British Isles when we speak of milk we generally mean cow's milk, though there are other countries where sheep or goats are the chief sources of milk. In peace time two-thirds of all milk produced in England is sold and used as milk; the rest is used for cheese making, for fresh cream, for condensed and dried milk, for ice-cream and other milk foods, and for butter making. Milk is a good food; it is easily digested because it is naturally produced to nourish growing young animals; it contains almost everything that is needed in a food, and is particularly suitable for children. This is why children are now encouraged to drink milk at school.

In recent years, for the sake of better milk, British farmers have been urged to pay great attention not only to the health of their cows but to the cleanliness of cowsheds, churns, and buckets, and the hands and clothes of the milkers. This improvement on dairy farms has produced better milk and has been followed by great efforts to persuade British people to drink more milk; and, in fact, more milk has been drunk as a result, but there is still much room for improvement. In England,[1] on an average, each person drinks about one-third of a pint daily; in America the average quantity is 1 pint; but in Switzerland, which holds the record, it is 1⅔ pints per person. Large quantities of butter and cheese were once made in the British Isles; but because prices were bad, less and less has lately been made; just before the war only one-tenth of the butter eaten in this country was made here, and not quite one-half of the cheese.

It is a strange thing, hard to explain, that in some parts of the British Isles very little fresh milk at all is bought, condensed milk in tins being used instead. Possibly some people like condensed milk better because it is sweetened, and is easier to buy and store. In warm weather fresh milk will not keep for more than a few hours, but milk from a tin does not go sour for several days.

Wool and leather. As everyone knows, sheep are kept for their wool as well as for their mutton, though not all wool from British sheep is suitable for making into clothes. Sheep that live in cold and exposed places grow a wool that is too coarse for clothes, and this is used for carpet making. The wool of all other sheep is suitable for clothes, the poorest or coarsest being used for suits, and the best or softest for underclothes. The finest kind of wool however does not come from British sheep at all,

[1] These three figures are for 1932, by 1939 the figure for England and Wales was 0·42 pint.

but from a breed of sheep called Merinos, which live in Australia and South Africa. Merinos are not as good for mutton as the sheep of British breeds.

Wool makes very good clothes; it wears well, and it keeps the wearer warm because it is a poor conductor of heat.

Sheep are usually shorn here between May and August, when the coldest weather has gone; they are not generally shorn until they are at least one year old (p. **33**). Of the wool used for the making of clothes in Britain nearly one-half is produced at home.

Leather for the making of shoes and gloves and clothes also comes from animals. The skin from cattle and horses is used for making stout leather for the soles, and the skins from sheep and from young cattle and horses, which make soft pliable leathers, are used for the softer 'uppers' of boots and shoes. These thinner leathers are also used for gloves, and for coats to keep out the cold winds. Not much leather comes from pigs, because when pigs are killed for bacon or pork the skin is generally left on the meat. Bacon rind is the skin of the pig.

Animals for work. Although tractors and lorries are now widely used on British farms many of our farmers still depend on horses for working their land. On a few British farms oxen are still used, as they were long ago, but they are not as good as horses for ordinary farm work. People particularly interested in engines may sometimes be heard to prophesy that working horses will disappear from all British farms in the near future; but that is a mistake. It is true that an ordinary tractor will do as much work as four or six horses—bigger tractors will even do the work of more than this—but many farms in this country are so small that all the work can be done by a few horses; certain kinds of work can best be done by horses anyhow, so the small farmer will therefore continue to keep horses rather than a tractor. On grassland farms there is little work that can be suitably done with a tractor, and much that can be done with horses; and on farms that grow fruit and vegetables, horses are still found to be more useful than tractors. For these reasons the working horse is not likely to go out of use on British farms.

The commonsense of it is, of course, that horses and tractors each have their special advantages. For hard work such as ploughing the tractor is very good, but the land must be dry and firm so that the tractor wheels can hold the ground and neither 'spin' nor sink too deeply.

The tractor does not get tired as horses do. If a farmer has been delayed with his ploughing he can catch up with a tractor, because he can work it for many hours each day. Sometimes tractors are worked all daylight and, if there is a moon, at night as well. Of course the tractor-driver gets tired nevertheless, and at such times the farmer will arrange for two men to drive the tractor in turn.

On big fields, and where the soil is light, the tractor is better than the horse for such work as drilling, rolling, and harrowing. Some farmers fix 'trains' behind the tractors; that is to say, one tractor will draw a drill, followed by several implements such as rolls and harrows. But such 'trains' can only be used where the field is large, because turning on the headlands is slow and difficult, and would be far too frequent in small fields.

Finally, whenever much produce or manure has to be taken long distances over roads between fields and farm buildings the tractor fitted with rubber tyres is better than horses, because it can take big loads quickly.

Horses are better than tractors for working small or odd-shaped fields; they can be used for hoeing roots and other light cultivations; when a farmer keeps live stock, and must feed them throughout the winter, horses can take hay, roots, and straw to the stock in buildings or on grassland, and can conveniently haul away the manure from yards and boxes; in short, on all the occasions when farm carts are needed, horses are likely to be preferred to tractors.

How live stock make good use of farm crops.

Grass is the largest single crop grown in the British Isles, yet grass as a crop is of no use at all to human beings. Cattle, sheep, horses, and goats make the best use of grass (and of foods made from it), but both pigs and poultry can also eat a little. The only way to make grassland useful to human beings is to keep on it animals that can be used for food and work.

In 1939 in England and Wales there were 23,388,947 acres of grassland, and in Scotland 13,543,146 acres. This includes not only the good fields with rich grasses and clover, but also the poor mountain grass that will only support sheep or goats. Of course, animals will eat grass as it grows in the fields; but in the winter there is not enough growing to satisfy them.

Grass may be preserved for feeding to stock in the winter in three different ways: as hay, as silage, and as dried grass.

Hay is made by cutting grass in the field, and letting it dry in the sun for several days. The time it takes to make hay depends upon the kind of grass, and upon the warmth and dryness of the weather.

Silage is made by cutting the grass and putting it while it is still green and full of sap into a special structure called a silo. There are many kinds of silos, made of wood, or concrete, or asbestos, or wire and paper. All are cylinders, and they may be anything from nine to twenty feet across and from eight to forty feet high. Silage as a food for stock has been more and more used of late years.

It has been found that the silage is improved if molasses (or treacle) is added to the grass at the time the silo is filled.

Dried grass as a food is an even more recent discovery. It is made by cutting the grass young and drying it by heat in a kind of oven. Of course dried grass costs more than hay or silage because of the expense of fuel for the ovens, but if good young grass is dried in this way it forms a food that is equal to some of the best cattle cake.

Many farmers grow crops like mangolds, turnips, swedes, marrow-stem kale and some kinds of cabbages to feed to cattle, and sheep, and to a lesser extent for feeding to other farm animals. In war time these foods are given in greater quantities than in peace time. It will be noticed that those crops which are known as cleaning crops are grown with two objects in view, the first to supply food for live stock and the second to give the farmer an opportunity of cleaning his land.

On the arable land, clovers and grasses, trefoils, sainfoins and lucerne are all grown individually, or in mixtures for stock feeding. These may be treated in the same ways as described for grassland in the above paragraphs.

Crops of peas, beans and oats are grown principally for feeding to live stock but, of course, special varieties are selected for feeding to stock and others for human food. These crops can be given to all farm animals, provided they receive proper quantities and ratios. A further advantage of growing peas and beans is that these crops have the power, common to all the pea family, of taking nitrogen from the air and in time enriching the soil they grow in.

By-products used for animal food. The main reason for growing cereals (wheat, oats, barley) is, of course, to get the grain; but the straw that remains after threshing is valuable too, and is not wasted. Straw is a by-product of every farm that grows corn. Straw can be sold for use in paper-making, or for packing; but the straw that remains on the farm is fed to animals or is used as bedding for them.

The best of the grain threshed out is sold for human food; but the small broken grains (amounting to 10 per cent of the total crop) are again a by-product, and are used as food for various farm animals.

When sugar-beet is grown, the roots are cut off and sent to the sugar-beet factory, where the sugar is extracted; the green tops are a by-product, and can be fed to cattle and sheep and, in limited amounts, to pigs. With such crops as potatoes or cabbages and brussels sprouts there is always a proportion that is not good enough to be sold for human food; the small potatoes are saved and fed to pigs or poultry, and the waste from green crops is fed to sheep and cattle. The stems or vines of peas grown for market as food make excellent food for cattle. They can be fed green, or as silage, or as dry straw. These by-products of the farm, which might otherwise be wasted and thrown away, are thus profitably saved and put to good use by the farmer who keeps live stock.

Even when the produce of a farm—say grain, or sugar-beet, or animals for meat, or milk—has been sold to be made into human food, parts of it become by-products in course of preparation and are bought back by farmers for food for their animals.

At the flour mills, bran and 'middlings' (or 'weatings') are made from those parts of the grain not used for flour. Bran is fed to all farm animals, often as a medicine; and the middlings are given mainly to pigs and poultry.

At the sugar-beet factories, after the sugar has been extracted from the root, a pulp remains. This is sold either wet or dried, as an animal food. Generally it is fed to cattle, but horses, sheep and pigs can also eat it, and so can goats and poultry, though only in smaller quantities.

Treacle, another by-product of sugar manufacture, is also fed to cattle, and is also used in making silage.

At the butter factory, skimmed milk is a by-product; and at the cheese factory, whey. The better qualities of these can be dried and used for human food, but the second-rate skimmed milk and whey are given to calves, young pigs, and poultry.

At the slaughter houses, where animals for meat are killed and the carcases prepared for the butchers' shops and so for the table, the by-products include dried blood, bone flour, meat meal, and meat-and-bone meal. The best qualities of each of these by-products can be made into foods for animals, chiefly the quick-growing animals such as pigs and poultry, and to high-yielding dairy cows. The by-products of slaughter houses not fit for foods are sold and used as manure.

All of the by-products so far named have had their origin on the farms; but there are industries which, while they do not get their produce from English farms, can nevertheless supply the farmer with by-products useful as food. The most important of these are the various oil cakes. In the manufacture of oils and fats certain seeds or nuts are crushed and the oil is pressed out of them. The meal that remains is compressed into cattle cake. Linseed, cotton-seed, earthnut, palm kernel, and soya beans, are all used in this way. These oil cakes are excellent for feeding to cattle and sheep, and, in limited quantities, to horses and pigs.

Waste from the fishing industry—scraps of fish and bones—is made into fish meal, which is a valuable food for pigs and poultry and for young growing animals such as calves and foals.

Animals enrich the land. By means of animals, then, we see that man is able to turn to his own good use not merely the grass crop which he grows for that purpose, but all kinds of products that would otherwise have to be treated as waste. But the waste products of the animals themselves are also of great value. As everyone knows, dung, which is what is rejected and dropped by the animal after food has been digested, is given back to the land as manure.

There are two ways in which the waste products of animals can be used to enrich the land: the animals may either be kept indoors (or in a yard) and the dung they produce carted out to the land; or they may be kept on the land itself so that the droppings enrich it where they fall.

When live stock are kept in buildings, that is in sheds or yards or 'boxes', the dung and urine will be mixed with the litter provided for bedding. Straw, bracken, sawdust, wood shavings, and peat moss are all used as litter; farmers believe that straw makes the best manure, and shavings and sawdust the poorest. The quality of the manure depends upon whether it comes from yards in the open, or from sheds under cover; because in the open it is washed by rain and the more soluble parts are washed away. Quality also varies according to the kind of animals; poultry manure is much richer than cattle manure, and manure from fattening cattle is richer than that from dairy cows.

Farmyard manure produces the best results in light soil, and gives the most profitable money returns from market-garden crops. On heavy soils, though manure is not essential as it is on light soils, it also brings good results; and even on Fen soils already rich in humus (which is the decayed remains of plants), farmyard manure produces a surprising improvement in crops.

In the other method by which animals are used to enrich the land, the animals live out of doors and the manure they leave behind falls directly to the land. For this purpose sheep have long been folded in hurdles on arable land, and in more recent years pigs and poultry have been used in the same way. During the last few years many farmers have been growing sown grass mixtures on their arable fields, thus turning them into pastures for the time being. These grasses are grazed by animals for several years, and then the enriched fields are ploughed up again. After the grazed land has been ploughed again crops can be grown on it for several years without applying further dressings of farmyard manures. This system of changing from arable to grass and from grass to arable is called 'alternate husbandry'. It is very successful on chalk soils (in the south of England), on heavy clay soils (in the Midlands and East Anglia), and also on medium soils (in the south and south-east of Scotland). In fact, it is successful on most kinds of soils and may be found practically anywhere.

SUMMARY OF THIS CHAPTER

Perhaps now we may better understand how much the farmer meant when he said that he expected his animals to keep him. This general chapter shows how essentially a part of the farm his live stock are. They are not merely producing—meat, milk, bacon, mutton, wool, leather, and eggs—they are a part of the working of the farm too. Animals help in labour on the land, they enrich the land with their manure, they make it possible to turn to human use the yearly crops of grass and hay, and many by-products (not only of farming but of other industries) which would otherwise be wasted.

Though the farmer must work to care for them and pay money to feed them, their lives are a necessary part of the goings-on of the farm, and they themselves, alive or dead, are also the sources for all of us of essential food and clothing.

CHAPTER II. GENERAL MANAGEMENT OF FARM ANIMALS

THIS chapter will deal with the best ways of feeding, housing, and breeding, and the general handling of live stock on the farm.

Feeding. It is useful to know and to remember that the feeding of farm animals and the feeding of human beings have much in common. The foods are different, naturally; humans do not eat grass, hay, and raw roots (except carrots), as cattle do; but animals have their likes and dislikes for certain kinds of food (as humans have theirs for greens or rice pudding), and a good stockman knows this and is ready to alter the foods accordingly. For instance, animals never like new foods at first, and therefore should not be forced to eat them in large quantities; they must have time to get used to them. Indeed, it is never wise to make sudden changes in the food given to farm stock; the change to the new food may upset them; and this is no more than common human sense, known to everyone who has, for instance, unwisely eaten too many apples early in the apple season. Very young animals have, however, even more than this in common with young human beings; they live upon milk as their first food, and the milk they either get naturally from their mothers by sucking, or are given it by hand. If milk is given artificially, it is important that it should be given at the right times and intervals, in just the right amounts, at the proper warmth, and out of clean vessels. Human babies when they are not suckled by their mothers are given milk from a bottle; when orphan lambs or piglings have to be reared by hand they also are fed from a bottle; young foals and calves, however, can usually be taught to drink milk from a bucket.

One other likeness between young stock and children may be mentioned: if the wrong foods are given to them, especially when they are very young, they may get a disease (called rickets) which causes a weakness in the bones, and is shown, for instance, by crooked legs. If this wrong feeding goes on, then the young humans or animals (or fowls, too) may grow up crooked and deformed; but if a change to the right food is made in good time, rickets may be cured and these early deformities may put themselves right.

In fact, animals and poultry must be fed with great care and knowledge, especially when they are young. The young stock must be given small quantities of good foods, such as good cakes or corn, good hay, and roots. Right feeding at this stage lays the foundation of later health. Then, as the animals get older, they will need less consideration and can be used as scavengers.

Scavengers largely find their own food from the land where they are quartered; they are generally called stores or store animals. Stores are expected to grow slowly, but they are sometimes fed so poorly that they are stunted and do not grow at all. Fifty years ago all stock was expected to have a long time as stores; but the more recent practice is to shorten the store period; and some animals—pigs, for example—nowadays are not kept as stores at all, but are fattened at once. Sometimes cattle and sheep may be treated in this way too.

Stock are fattened at various stages in their lives. Some are fattened when very young, others only when they are mature (or full grown). Fattening is done by liberal feeding, but the foods must be good, and to entice them to eat freely they must be tasty. Liberal feeding is essential, but overfeeding is, of course, a mistake to be avoided;

overfed animals will not thrive, and the food will be wasted. Stock will fatten best if they are fed at frequent intervals. Where store stock may be fed twice daily, fattening stock may be fed three or four, or even sometimes five, times a day.

Many farm animals are kept for breeding. During the breeding season the males may need extra foods; but at other times they will be fed as store animals. The females need very little extra food until shortly before the young are expected. Then extra food is generally necessary, because the mother, or dam, will now have to feed herself, and produce sufficient milk at the same time for her progeny.

The milk of some farm animals is richer than that of others; for instance, sow's milk is richer than cow's milk. Some animals give more milk than others, the dairy cow, for example, gives much more than the heifer. Feeding will vary accordingly: animals giving rich milk must have rich food with which to make the milk, and those giving big quantities must have more foodstuffs than those giving small quantities. Piglings grow very quickly; they weigh often only 3 lb. when they are born and yet at eight weeks old they will usually weigh 30 lb. apiece, ten times their weight at birth. Calves weigh 70 or 80 lb. when born, but it takes most of them one and a half years to two years to reach ten times their birth weights. The rapid growth of young pigs is largely from the sow's milk; it is clear therefore that sow's milk is rich in minerals and proteins, which are known to be essential for body growth. Whenever animal mothers are producing milk they must be fed according to the quantity and quality of milk being produced.

The foods that are given to farm animals may be divided into three main groups: the rich foods called ' concentrates '; the bulky fibrous foods called ' roughages '; and the bulky moist foods called ' succulents '.

The concentrated foods, cereal grains, peas and beans, and purchased foods such as cakes and meals, contain much real food or nutriment in proportion to their weights. These concentrated foods form the daily diet of pigs and poultry because neither the pig nor the hen can digest much of anything else. Concentrated foods are given in limited quantities to all other live stock, except when they are grazing in spring or when they are passing through a store period.

The roughages are the hays and straws. These are given to cattle, sheep, and horses, because they can digest bulky foods; in fact, they need them, and if they are not given enough to make them contented, they will eat the woodwork of fences or partitions. The good hay is kept for the young animals, for those producing milk or fattening or working; and hay of poorer quality and straw are given to the store stock. It is really true to say that pigs and poultry eat no fibrous roughages at all.

The succulent foods are those that naturally contain a high percentage of water, such as roots, cabbages, grass and silage, though very good grass and silage may almost be reckoned as concentrated foods. These succulent foods all have a laxative effect on the animals eating them. In moderation this may be a good thing for them, but if large quantities of succulents are given some of the dry fibrous foods (roughages) may be required to correct this tendency.

All farm live stock eat grass, and they will also eat a certain amount of the other succulent foods, but cattle and sheep will eat more of them than other animals; and of the cattle and sheep it is those that are being fattened that will eat most of the roots. These succulent foods have slight medicinal properties, for animals that are slightly ill may be coaxed to eat more freely by offering small quantities of succulents, as human patients may be given oranges, or apples, or grapes when they are ill.

It is often necessary to prepare foods by crushing, grinding, chopping, or soaking them, before animals eat them:

(*a*) Cereals are crushed to make sure that if any grains are not chewed by the stock eating them they are still exposed to digestive juices, and may be digested. Cereals and beans are crushed before feeding to cattle, sheep, and horses.

(*b*) Cereals and peas and beans are ground into meals for feeding to pigs and poultry.

(*c*) Sometimes cattle cakes are cracked to make them smaller, so that stock may eat them with safety.

(*d*) During the last few years concentrated foods have been made into cubes or pellets for all farm stock. The stock prefer the food in this form rather than as meal.

(*e*) Hay and straw is often cut into short lengths by means of a chaff-cutter. The chaffed food is fed to horses, and sometimes to sheep and cattle.

(*f*) Mangolds and swedes are sliced to entice cattle and sheep to eat more of them than otherwise they would.

(*g*) Various meals are soaked in water for a few hours before they are given to live stock. This is done to make digestion easier. Foods are soaked for feeding to pigs and poultry. Dried sugar-beet pulp is soaked before it is fed to cattle.

(*h*) Some foods are cooked before they are fed to stock. It rarely pays to cook foods for live stock except for pigs and poultry. Of all the foods that are cooked potatoes are the most popular. By law, household waste food must be cooked before it is fed to pigs. This is to reduce the risk of spreading diseases.

Water for drinking must of course be provided for all stock. The amount they will drink depends upon the weather, the kind of animal, and the foodstuffs being consumed, and if much wet food is being given then less

3. Fattening cattle in an open yard

4. Mountain sheep at home

water will be drunk. Most water is needed in hot weather; cows in milk drink more than fattening bullocks. Hens will not lay eggs and cows will not give milk if they are given too little water. Animals die more quickly from lack of water than from lack of food. Water is essential for all farm animals.

Housing of stock. In England some stock are housed, or kept inside, all the year round; others are housed in the winter only; and some are never housed at all. Most sheep are never housed, though certain breeds may be put into pens at lambing time (p. 32). Many breeds of beef cattle, too, live entirely out of doors, and may even bear and rear their calves in the open. At the other extreme, certain stock that are housed throughout the year are treated so because they are tender and cannot stand cold weather, or because they are worried by flies in the summer. Calves from dairy herds are often kept indoors till they are at least six months old.

Cattle may spend their fattening periods indoors (see picture opposite), or on grassland. Indoors they cannot take exercise, and will therefore get fat more quickly. The fattening of pigs, at any time of year, is almost always an indoor process (p. 64). In some systems of poultry keeping, too, chickens spend their whole lives indoors: where they are hatched, reared, and either fattened or kept for egg production (pp. 80, 81). Some readers may be surprised to hear that in some towns cows are kept entirely in dairies. These cows live in cowsheds in the town itself. The advantage is that the milk does not have to be taken great distances to the consumers. In Liverpool, which is the only town where dairying is still carried on largely, the city cowsheds are very good—much better than in many rural areas.

Buildings for stock must be warm, dry, well-ventilated and free from draughts. If the buildings are not warm enough the animals will need more food to keep themselves warm. Dry buildings, besides being naturally warmer than damp buildings, will need less straw to make dry, comfortable beds for the stock; this means that the animals will lie down more, and will as a result fatten more quickly. If the bedding is damp they will stand and walk about and be restless. Thus there should be a good roof against the rain, and good drainage to the floor.

Fresh air is good, and draughts are very bad, alike for live stock and human beings. Draughts produce a body chill and make stock, or human beings, catch colds. For various reasons it is equally bad for stock (and for us) to live enclosed with too little ventilation; in particular, certain air-borne diseases spread most rapidly in badly ventilated buildings. Besides, if animals are kept under bad conditions they will not be fit enough to resist disease if it should appear. Bad ventilation may thus mean that sicknesses will continue longer and affect more animals. Special care is usually taken to see that stock kept for breeding shall live in well-ventilated buildings and under healthy conditions. Many farms have special buildings for certain animals. Working horses are usually kept in stables; but in arable districts, where straw is plentiful, they are often kept in yards either partly or completely under cover. In a yard the horses are able to get exercise, but there is just the chance that a bad-tempered horse may start to kick others, and serious trouble in the yard may follow. All young horses live out of doors in summer, and some may remain outside during the winter too. In cold districts they may either be kept in yards in winter, or if they remain out in a field, there will be a shed into which they may go in bad weather.

Formerly pigs were kept in shockingly wet, dark, and dirty places, but during the last twenty years farmers have learnt by experience that pigs thrive best of all if kept in warm, clean buildings. There has lately been a growing tendency to keep breeding pigs on pastures, and on some farms sows are kept on grassland in portable huts that are moved to fresh grass almost daily (p. 33). For fattening, pigs are either housed in boxes (four to fifteen per box), in the old-fashioned sties, in the yards that are used at times for cattle, or in a 'Danish piggery' (p. 64) —an imperfect copy of buildings specially designed for pig-keeping in Denmark—a long shed covering two rows of pens. Between the rows of pens is a passage called a 'feeding passage' down which the pig's food is taken. Down the outside of each range of pens runs another passage called a 'dunging passage' in which the pigs leave their dung and urine. The piggeries in Denmark (but frequently not those in England) have an upper storey, filled with straw. This keeps the whole building warmer in winter, and the pigs fatten more quickly accordingly.

Calves are often kept in small pens or boxes. They catch cold very easily in cold and draughty buildings, and above all it is essential to make sure that the beds of straw on which they lie are warm and dry. After their third month calves thrive best of all in yards.

For fattening cattle, yards (p. 17) and boxes are commonly used, the boxes being best when the animals are almost fat.

Dairy cows in winter may live entirely in a shed where they can be fed and milked; or they may be kept in a box. On some farms the cows are kept outside except at milking time, when they are brought into a special milking-shed. If a milking-shed (p. 16) is used the cows may also spend the time between milkings in yards. During the last few years portable milking-sheds have been made that can be taken to the cows in the fields. These sheds are used on light soil throughout the year, but on heavy soil in the summer only, because the cows would poach heavy land in winter. In some districts cows are milked in the open fields in summer without buildings of any kind.

It will be seen that boxes and yards may be used from time to time by all farm animals. Boxes can be most useful for any sick animal, or for an animal that is kept apart for her young to be born.

Various special buildings are used if poultry rearing is done on a big scale. There are the incubator rooms, followed by the brooder houses (p. 80), which must be provided with artificial heat. When chickens are two months old there are Sussex Arks for birds kept outside. These are like large dog kennels. Sometimes at this age the birds are kept indoors entirely, five birds to each pen. When birds are laying they may be kept indoors also (one to a pen) in what are called 'batteries' (p. 81), or they may be in movable laying houses on pastures or stubble (p. 80). Other poultry farmers keep the birds in large fixed houses with large permanent grass runs. Laying houses are fitted with nesting boxes and roosting racks, and generally also with troughs for food and water. Poultry buildings are usually of wood, and must therefore be treated with creosote to keep them free from the parasites that live on the birds.

Breeding.   Stock chosen for breeding will be strong and healthy and fully grown. Breeding from stock not fully grown may make both the mothers and their offspring stunted. When the farmer knows that they are mature enough he will begin to breed from them. He usually arranges that the young shall be born at a time when he will have enough food for the mothers and their progeny. Breeding begins when the male deposits sperms within the female, who in her body provides minute 'eggs'. When the sperm from the male meets and unites with the eggs in the body of the mother the eggs have been fertilized, and while the young animal is formed from the fertile egg within its mother's body a period called the gestation period passes before the young are born. The gestation period is therefore the time between fertilization by the male and the birth of the offspring; this time is different for the various farm animals; for a sow it is four months; for a ewe, five months; for a cow, nine months; and for a mare, eleven months. Farmers of course must know these times, and they should keep records of the fertilization of the females (known as 'service') so that they can make proper preparations at the time when the young are expected to be born.

The age at which animals are ready for breeding varies with the kind of animals. Ewes are usually two years old when they have their first lambs; gilts (young sows) are one and a half years old when they have their first litters; and heifers (young cows) are two to four years old when they have their first calves. These ages, however, differ slightly with different breeds of the same animals; some cattle that are kept in cold, exposed districts are slow to mature for breeding. Fillies (young mares) are three or four years old when their first foals are born. Ewes and mares give birth to their young in the spring when the weather is warm and there is plenty of grass for them to eat. It is possible for cows and sows to have young at any time of the year, though often, by controlling the time of service by the male, many farmers make sure that the beef cows have their young in the spring.

Males are first used for breeding at the following ages: ram lambs at seven months, boars at twelve months, bulls at fifteen months, and stallions at two or three years. If males are used for breeding before they are fully grown they may become stunted, or they may lose their power of fertilizing the female, when they are said to be sterile and are of course useless for further breeding.

The male may serve many females, and he is the father of many animals each season. Each female, of course, has only the young she carries during gestation: the mare one in a season; the cow one, and occasionally two; the ewe one, two, and occasionally three; and the sow two litters of eight piglings in a year. Because one sire can become the parent of many offspring in a season a farmer who wishes to improve his stock can do so most rapidly by using good males for breeding. Therefore whenever a farmer buys a new male, or sire, he should take the greatest pains to choose a good animal—good in appearance, and one that is likely to leave good progeny. It is impossible to be sure that a sire will have good progeny; but one that comes from stock known to be good is thought to be more likely to leave good progeny than a sire from an unknown source. Money spent in buying a good sire is well spent. Farmers say a good sire is half the herd, stud or flock. Undoubtedly it pays best to keep good stock; bad stock need just as much care to look after them, they eat the same food; yet good stock make better use of the food, and leave better progeny.

Usually good stock are 'pedigree' stock, which means that their ancestors are known to have been good, and have been recorded in a register of stock of that breed. Such a register is called a herd book. Each breed of stock has its own herd book. Good pedigree British stock sell for very high prices, because they are much sought after by breeders of live stock in this country, and sometimes also by breeders in other countries. Because there are so many breeds of good stock in this country the British Isles are sometimes called the Stud Farm of the world. The production of first-class stock depends on good breeding and skilful rearing and feeding. The British farmer is renowned as a skilful breeder, and British stockmen are recognized as first class for rearing and developing the stock for sale. In normal times live stock are exported from Britain to all parts of the world, and this is likely to go on as long as the British reputation for good stockbreeding is maintained.

Handling of live stock. From time to time stock may be driven or led. Good stockmen always treat animals with great care; for stock badly treated quickly become nervous, and once animals are nervous they never thrive really well. It is wise to see that stock are handled quietly without being beaten and frightened. Many animals are very strong, and if they are treated quietly they never need discover their strength. If a bull is put into a shed, a yard, or a field he must be encouraged to stay where the farmer wants him. Once he has learnt the way to get out he may break out frequently. The bull is one of the strongest of farm animals and this must be remembered in deciding where he is to be kept. Where stock are kept on pastures they are most likely to get out when the supply of food is limited, or in the breeding season.

Animals soon become used to a routine and they quickly notice changes and become upset by them. If they are not fed and milked at regular times they will grow restless; and restlessness is bad for them, they will not thrive and the food they eat is wasted. A stockman who keeps to a strict routine produces better stock than one who looks after his stock haphazardly.

When animals are unwell they lose their appetites, just as boys and girls do, and if any animal does not eat when food is provided, then the farmer or stockman should try to ascertain the cause. The dung should be of the proper constituency according to the animal, e.g. that from a horse or sheep is much firmer than that from a cow or pig. Illness is clearly shown by abnormal dung being produced. The healthy animal is alert and watchful, whereas a sick animal is sleepy and dejected. When stock are out on grassland, they keep together as a flock or herd. Any animal that is isolated should be viewed with suspicion. A female that is about to have young will often hide away from the other stock. Unless an animal has taken a lot of exercise it breathes normally and does not perspire, thus whenever there is panting and perspiration for no apparent reason illness may be suspected. Abnormal walking or lying should be viewed with suspicion, since it may be more serious than just injury causing lameness.

A good stockman is able to see that an animal is unwell, and to give it special attention, almost before it feels ill. Early treatment like this often reduces the severity of illness and sometimes avoids it altogether. It is quite evident that much of the success of keeping live stock depends upon the skill of a careful, painstaking stockman.

Grooming is another duty that no good stockman will neglect. It is done to remove dust, scurf, and loose hair from the animal's coat. Sometimes the coat needs to be clipped either (with horses) to make grooming easier, or (with cows) to make sure of clean milk production. Some cattle must have their horns trimmed because they get long and dangerous; and the hoofs of all farm animals need trimming or paring from time to time. The stockmen in charge of the stock should be able to do all such jobs as these. Any good stockman knows quite a lot about the practical work of feeding and breeding and may need very little supervision, beyond encouragement from his employer. There should be no lack of encouragement; undoubtedly a good stockman is worth his weight in gold.

CHAPTER III. FROM CALF TO DAIRY COW

In a wild state calves are naturally born in the spring, but the domestication of cattle has changed this wild habit in some degree. Calves for dairy herds may be born throughout the year, though perhaps rather more are born in the spring than at any other time.

The time of year in which a calf is born is important, because the method of rearing will be varied according to the weather of the first few months of its life. We shall describe the rearing in summer and in winter till the calf is one year old. After this the next stage, of one to two years, will be described; and finally the treatment of animals of over two years.

The first year. A calf may either be taken away from its dam as soon as it is born, or left with her for several days, or it may remain till it is weaned. The calf that is left with its mother till weaned (that is, till it needs no more milk) will make the quickest growth, but this is expensive because the cow rears only one or, at most, only a few other (foster) calves each year. Such suckling calves grow best because by sucking they get the milk at exactly the right temperature, and take only small quantities at a time; also the milk is much freer from dirt and the germs of disease than if it is milked from the cow and exposed to the atmosphere before the calf gets it. A cow may suckle her own calf alone (which is very expensive because the one calf really has to pay the full cost of keeping the cow for a year), or the cow (as foster mother) may suckle several calves. It is usual for a calf to feed from a cow for two or three months. A good foster mother will suckle four at a time (p. **16**); and in a single lactation period of about ten months' during which she continues

to give milk after a calf has been born, she may rear eight
and sometimes even twelve calves. Some cows will not
allow calves (sometimes even their own calves) to suck;
others are willing to feed any calf.

In the spring and summer a cow and her own calf may
be kept out on grassland together, but in winter the calf
generally remains indoors. When a cow is rearing several
calves it is necessary to keep the calves in buildings, though
the cow may go outside throughout the whole year. When
a foster mother comes in at suckling time (night and
morning) the stockman ties up the cow and lets the calf
or calves suck. It may be necessary for him to be there
to see that each calf gets its fair share of milk, because
sometimes the biggest calves will take too much, and
leave too little for the small calves that need it more.

Much more commonly calves, instead of being suckled
by a cow, are reared on milk given to them in a bucket.
More calves can be reared upon the milk of one cow by
this method; but the successful growth of such hand-fed
calves depends entirely upon the skill and care of the
stockman. Each calf must have just the right quantity
of fresh milk from clean pails, given at the proper times
each day, and at the right temperature. Sometimes when
calves are a few weeks old they are given substitutes for
cow's new milk, such as gruels made with water, skimmed
milk, whey, and moistened dried skimmed-milk powder.
Any of these substitutes, if given with care and knowledge, can produce good results.

No matter upon what system a calf is to be reared, it
always makes the best growth if it has milk from its own
dam for the first few days of its life. This first milk, given
by an animal after its baby is born, is called colostrum,
and is quite different from other milk. It is deep yellow
in colour, richer than ordinary milk (it contains more
proteins, and is equivalent to a mixture of eggs and

ordinary milk) and has a laxative effect upon the young animal. It gives the newly born calf a good start in life.

When calves are about three weeks old they will begin to nibble such solid foods as crushed oats, flaked maize, whole beans, and linseed cake. These should therefore be provided for them in boxes or troughs. At the same time they will also begin to eat hay. They should be given good well-made hay, such as that from lucerne, meadow, or clover. As soon as they begin to eat these dry foods they will get thirsty, and water should accordingly be there for them to drink whenever they want it. Calves also like to lick a lump of salt now and then at the time when they begin to eat solid foods. As soon as a calf is eating solid food the quantity of milk or milk-substitutes[1] fed may be reduced; and by the time that it is three months old it may be fed entirely upon hay, concentrated foods, and water. No milk is given to calves of this age intended for dairy herds, except occasionally to one that is backward.

Between the ages of three and six months the daily quantities of hay and concentrated foods will be increased, and roots may be introduced into the ration. Silage is not usually fed to calves before they are six months old. There is a general agreement also that calves by themselves do not thrive if they are turned out to grassland before they are six months old. But it is a remarkable fact that if a calf is turned out to grass along with a cow and is still getting milk from the cow, then grass seems to be good, and the youngster will thrive. Grass is inclined to make calves scour; and on pasture in hot weather flies are always rather troublesome. It may be because of this that (especially in hot weather) calves do better when kept indoors.

[1] As a war-time measure gruel feeding was recommended, since a calf could thus be reared on 20 gallons of milk instead of on 80 gallons.

The proper treatment for calves between six months and one year will differ with the season during which they reach this age. Spring-born calves, which reach this age in autumn and winter, will therefore be kept indoors, and not turned out until the following spring. During that winter their ration will change from that given at six months of age to a store ration; the quantity of concentrates fed will be much reduced; the quality of the hay will also be reduced, though the *quantity* may be increased; straw may also be fed to them for the first time; and the quantity of such foods as roots and silage may be increased. During this winter period the most satisfactory way of keeping these yearling heifers is in yards either completely or partly roofed. If the yards have a southern aspect and are well sheltered on the north side they will be warm, and store cattle will then thrive well on relatively poor rations.

Autumn-born calves that are six months old by the spring may be turned out to grass in the south of England by the middle of May. But because the change from winter to summer feeding is very great, and calves six months old are really rather delicate creatures still, certain precautions must be taken. For the first few days they should be outside each day for a few hours only, a warm sunny day being chosen for the very first day out. They should be brought in and housed at night. The pastures chosen should have been fairly heavily grazed already so that they are unable to eat too much young grass. After a few days the length of time outside may be increased, until at length they are out all day. If there are heavy dews, however, or on wet days, it is best to keep them inside, because in some districts young stock may develop a cough (called husk) due to parasites living on wet grass.

During the first warm night they may be left outside, but should always be brought in if there is any danger that the night may grow cold. For such watchful care as this it is clear that they must be kept on grass fields not too far away from buildings. Some farmers keep young calves in orchards; they cannot do much damage to trees at such an age, and on sunny days orchards provide plenty of shade. When at length heifers are able to stay out altogether they are given nothing (except water) in addition to the grass they eat. But as the summer advances the grass may become scarce, through drought, and it may become necessary to give some food in addition to the grass. The food must be freshly cut lucerne or oats and tares, or cabbages or kale, or even green maize. By the middle of August it is necessary to watch for the first signs of coughing from the parasites producing husk. As soon as any cough is noticed cattle should be brought in and kept in at night, and only allowed out to graze till the end of September; after that date yearling heifers under one year should be kept entirely inside.

The store period, one to two years old. Except for the Jerseys, which are exceptional and may have their first calves at twenty-one months, most heifers between the ages of one and two years are kept as store cattle. Whatever part of this period coincides with the summer will see the heifers out grazing, where they will receive nothing but grass. It will be necessary to watch them carefully when they are first turned out in the spring lest they scour too much, or eat so much grass that they become 'blown'. If they eat a lot of young grass quickly it will ferment rapidly in the stomach, forming gas which may so blow up the stomach that it presses on the lungs; in extreme cases the animals may even die of suffocation because of this pressure.

Heifers in store condition may be placed on second-rate pastures or on some of the inferior marshes; they can also be used as scavengers to clear up superfluous grass on certain fields or even on parts of fields. As scavengers they often follow behind dairy cows. The cows are allowed to eat the best grass, and are then moved on; because if they are made to eat the grass too closely their milk production may be reduced. Heifers may suffer from husk in the same way as the younger cattle (see p. 28), but are a little less likely to get it.

In winter time in most districts of Britain one- to two-year-old heifers are taken indoors, and kept in covered (p. 17) or open yards, or in boxes. Yards are better; cattle are generally healthier in yards than in boxes because yards afford them plenty of fresh air and the chance to take exercise. In the winter the heifers feed principally upon straw and roots, perhaps with silage and (occasionally) concentrated foods added. They must have plenty of water, and salt should be available to be licked as they want it.

Since many of the heifers are to have calves by the time that they are two and a half years old, mating or service by the bull should occur when the heifer is about twenty-one months of age. Usually, to make sure that breeding has started by the proper time, bulls are mated with the heifers when they are eighteen months old. If heifers are living out on pastures the simplest way of breeding is to keep a bull constantly with them. The disadvantage of this is, however, that the actual date of service by the bull may not be known; and it is important to know this date if the farmer wishes to prepare the heifer before the calf is born. Alternatively the heifers may be watched carefully each day to see when one or the other is ready to breed. This readiness to breed is known as being 'on heat', and is shown by a heifer when she jumps upon the backs of others and calls to the bull. Heifers seen to be ready to breed are then taken to a bull that may be

kept elsewhere. By this means the farmer is able to keep
a record of the exact date of service. When heifers are
kept in yards a bull may run with them as in the pastures,
but, as before, no exact records can then be kept of service. In boxes bulls cannot be kept with the heifers
at all.

The stockman must watch the heifers for symptoms
that they are on heat, and should take them to the bull
as soon as they are found to be ready. No change is
made in the feeding or housing of the heifers when they
start breeding, they are merely watched closely to see
if they have been successfully served, for if they have
not they must visit a bull again. Occasionally a heifer
will not breed, and then she must be fattened (see
Chapter IV).

Over two years: the heifer's first calf. The change
in feeding and treatment of a heifer over two years old
is not regulated by her age, but by her breeding. The small
calf growing within her (known as the foetus or the foetal
calf) develops for six months before it needs very much
food from the heifer, but for the last three months of its
foetal life it grows rapidly, and during these three months
the heifer must be watched carefully to see that she does
not become thin. She should, in fact, become slightly
fat during this time. Sometimes the heifer in calf will
be kept with the milking cows so that she may feed with
them, but there is always the danger that a heifer may
be bullied and hurt by the cows. When several in-calf
heifers are expected to calve at about the same time, it
is probably best to keep them by themselves, away from
the cows; but wherever they are kept they must be fed
as cows are fed. If these last three months happen to be
winter months, when the cows are kept in yards, then
these heifers will also live in yards. If the cows are

brought in to lie in sheds at night, the in-calf heifers
should do likewise. Similarly they should be fed the
same sort of rations as cows producing small quantities
of milk, that is to say with second quality roots and straw
(and possibly silage as well). If they are found to be in
rather poor condition, concentrated foods (mixed as for
the dairy cows) must be given to build them up. This
'building up' is not so much for the birth of the calf as
to prepare the heifer for the milk production that naturally
follows the birth of the calf. The quantity of concentrated
foods given each day may be as high as 8 or 10 lb. to
each heifer.

In summer time the in-calf heifers *and* dry cows (cows
producing no milk and due to calve shortly) are all kept
grazing together. These breeding animals (like the younger
heifers) may be used as scavengers after the milking cows
have had the best grass, or they may be grazed in fields
that are too far from the homestead to be used by the
dairy cows, which ought not to walk long distances to
and from the buildings for milking. If, through drought,
the supply of grass should get poor, it may be necessary
to give some additional food on the pastures such as
green maize, cabbages, kale, lucerne, vetches and oats,
or silage. If these foods, freshly cut, are given on the
pastures such feeding is known as 'green soiling'. Sometimes the in-calf heifers are given concentrates while at
pasture; these foods will be put out once or twice daily
in wooden boxes or metal tubs. This treatment is continued till calving is expected; but as calving time approaches the stockman should see the heifers at least
once or twice daily.

When the calf is about to be born, the heifer's udder
noticeably grows larger, and there are other obvious signs
that the calf is about to leave the heifer. In the summer
some farmers will leave heifers outside on pasture for the

5. Sheep dipping

6. Arable-land sheep in a lambing pen or yard

7. Shearing sheep by hand

8. Portable pig huts for sows and their litters

calves to be born, and in winter will let the heifers calve in yards or boxes, just wherever they happen to be. But those who take more care of their stock usually make sure that cows, and especially heifers, are put into special boxes or pens for calving, wisely arranging for the heifers to be watched by the stockman so that assistance can be given at calving if it is needed. Help given by the stockman at the right time, and in the right way, may save the heifer much pain; indeed, it may save the life of the calf, and of the heifer too. Sometimes the calf is likely to be smothered at birth by the foetal fluid, and only prompt action by a good stockman will save its life.

Directly the calf is born the heifer will turn to it and lick it dry. This process not only massages the calf, but it is known that the fluid licked up by the heifer has medicinal properties for her also. Now and then a heifer will be found to take no interest in her calf; a stockman must then dry the calf himself with wisps of straw, remembering that the drying massages the calf at the same time.

As soon as the calf is strong enough it will stagger up and begin to suck. If it should happen to be too weak to suck by itself, it may be lifted up and milk may be squirted into its mouth. Sometimes the heifer has more milk than the newborn calf can drink; if so it is essential to milk the heifer by hand. The stockman must take care, however, for there is a risk that too much milk may be taken away, and this seems to make her liable to get milk fever. Though nowadays cows or heifers can be cured of milk fever, formerly it was often fatal. To avoid the risk of fever the cowman should take milk from the heifer at frequent intervals. Within twenty-four hours after the calf has been born the 'after-birth' (the envelope in which the foetus or young calf has lived inside the heifer) should pass away from the heifer. Good stockmen should watch for this. In the wild state it seems to have been an instinct with a cow to eat the after-birth; it might be an advantage that thus no remains were left to be found by beasts of prey, to whom a female with her newborn young might fall an easy victim. This instinct seems to survive in some domestic cattle, for when domestic heifers are left to calve unattended they sometimes eat the after-birth.

Just after she has calved the heifer needs to be fed carefully; the diet must be laxative, and care must be taken not to give too much milk-producing food. Good hay and roots should be given, or good grass in summer, for maintenance, and the concentrated foods only according to the quantity of milk produced. Then it is necessary to start milking the heifer at the proper hours of milking the herd.

The milking herd. About four days after the calf is born it must be taken away and its mother put into the milking herd. This separation usually upsets both heifer and calf. The calf has to be taught to drink milk from either another cow (a foster mother) or from a bucket; and the heifer usually 'holds up' her milk, and she bellows and tries to get to her calf, so for a day or two she may be difficult to milk.

Milking must be carefully done if good results are to be obtained. It is done by squeezing the teats rhythmically, not by pulling on them. Speed is very important, because quick milking leads to increased milk yields. It is also essential to milk thoroughly, and to obtain the last drop of milk from the udder, and this for two reasons: the last milk is the richest in butter-fat; if milk is left in the udder the heifer or cow ceases to give milk and quickly becomes dry. Experience shows also that milking should be done according to a time-table at constant times each day; milking at irregular times will reduce the quantity of milk given. Cows should be milked twice or, if, milking well, three times daily, the greater frequency of milking producing more milk. The machine milking is good because it copies the sucking action of the calf, it also milks all four teats at once, and speed thus obtained results in a good yield.

Certain foods may cause a *taint* in the milk of cows that eat them. Turnips, cabbages, kale, lucerne, camomile, wild garlic, will all cause a taint in this way, and should not be fed to cows in milk; though there is less risk of producing taints if such foods are given just after milking. Milk may also be tainted after it has left the cow; by the smell of silage in a cowshed, for example.

During the last twenty years much attention has been given to the production of clean milk (p. 16). Cows, cowmen, buildings, buckets, churns, are all washed and cleaned daily, and preferably just before each milking. It may seem rather cruel to wash cows with cold water in winter time, but once they are used to this treatment no harm comes to them. Another important part of clean milk production is that milk must be cooled down to 50° F. as soon as it has been drawn from the cows. On some farms the cooled milk is put direct into bottles ready for the consumers. The sale of milk in bottles has taught housewives to look for the cream line, and farmers in their turn have thus found it essential to keep cows that produce milk containing much cream.

When a heifer is settled in milk she should continue milking for a period of ten months (i.e. for a lactation), and then should have a rest of two months before her next calf is born. This time of rest is called a dry period, and is necessary to let her recover from the strain of producing milk. Usually a heifer breeds (is found to be 'on heat') about three months after she has calved; if she is served then by the bull she will carry her new calf for nine months, which means that she will calve again twelve months after her first calf was born. When a heifer calves for a second time she is no longer called a heifer but a cow. (A heifer of one of the most common breeds should produce 600–800 gallons of milk in a lactation (ten months); a cow at maturity (third lactation and subsequently) will produce 1000 gallons of milk.)

A cow continues to breed for several years, but on an average remains in dairy herds for not more than three and a half years. There are exceptions however; many cows have been known to remain in the dairy herd for ten years, and there are records of even longer periods, up to twenty years, though these are, of course, rare.

Milk and its uses. Normal milk contains about 87 per cent of water. The remaining 13 per cent consists of fat, proteins, sugar and various other things. The proportion of fat varies chiefly with the breed of cow: for example, the Jersey often produces milk containing 5 or 6 per cent of butter-fat, whereas the milk from the Shorthorn contains from 3·8 to 4 per cent. The quality or richness of milk produced by a cow is not constant: it is higher when she is young, and also at times when the *quantity* she is giving is low. Generally the evening milk contains more butter-fat than milk at any other time in the day.

Milk may be sold as fresh milk or as pasteurized milk direct to the consumer for household use, or it may be sold to creameries for manufacture there into cheese or butter. Formerly butter and cheese were made on many farms, but the application of science to the traditional methods has made a great difference, and better results are now obtained by making butter and cheese under factory conditions rather than on farms. One gallon of average milk makes 1 lb. of cheese, but 2½ gallons is necessary to provide the cream for 1 lb. of butter.

Cream is sold by some farmers either to creameries or to the consumers. The cream trade is mainly confined to the summer and especially to the period when fresh fruit is available.

CHAPTER IV. CATTLE FOR THE MEAT MARKET

IF cattle are to be used for milk production they will be of the most suitable breeds for that purpose. But not all cattle are intended for milk production; some are intended from the first for beef, and some breeds are more suited than others to this purpose. Thus there are dairy cattle and beef cattle; and a third class that, because they are equally suited to either according to feeding and management, are known as dual-purpose breeds.

Five classes of butchers' meat from cattle are recognized: veal, baby beef, beeflings, mature beef, and cow beef. Veal is the meat of calves not more than three months old. Calves killed for veal are from the dairy and dual-purpose cattle; those from the beef cattle will not of course be killed so young but will be kept for beef. Most of the calves fattened for veal will be males, which are not wanted in numbers for breeding as the females will be.

Baby beef is the flesh of fat cattle killed at fifteen months old. Most if not all of these will be dual-purpose cattle; and again more males than females will be killed at this age. For this and the subsequent kinds of beef the bull calves are castrated when about three months old.

Beeflings are two-year-old fat cattle; they are either those from the baby-beef group that did not respond to the fattening treatment at that age, but grew instead; or they are those that, though intended for mature beef, were found to have fattened earlier than expected. Beeflings will thus all be from the beef and dual-purpose herds. Mature beef is from animals two and a half years old or more, chiefly from the beef or dual-purpose breeds; they will be of both sexes, though the majority will be bullocks. Cattle that are the offspring of dairy cows mated with beef or dual-purpose bulls will also often be kept and fattened for mature beef.

Cow beef is the meat of cows that, having ceased to breed, have been fattened for market. Beef cows and dual-purpose cows will usually fatten easily enough, but some of the dairy breeds cannot be fattened.

Different methods of rearing and fattening the cattle must be used according to the class of meat they are to produce, and the rest of this chapter will describe these methods more in detail.

Veal. The calves for veal are fed on milk, and to get the very best quality of meat no other food should be given. To make the calves get fat quickly they are usually kept in pens, one calf per pen. Whether in boxes or pens they are always kept in the warm and in the dark. To discourage exercise the calves are made so comfortable that they spend a lot of time lying down. Milk is fed to them from a bucket, and to encourage quick fattening they are fed three times a day from birth. They are usually fat and ready for market by the time they are three months old; at this age, if they have been fed on milk alone, their flesh will be white, like that of chicken.

It may be that the pure milk that can be spared for the calves is not enough to feed them entirely; in that case they may be fed on the same foods as calves that are being reared for the dairy herds—hay and roots and concentrates. They may also be given substitutes for pure milk such as gruels or skimmed milk. It is a strange fact that these foods produce flesh that is not white. French farmers believe that adding chalk to the foods will produce white flesh, but this is not true. Veal is very popular in the spring, and prices are usually higher then than at any other time of the year.

Baby beef. Calves intended for baby beef must be fed on fattening foods all the time from birth until they are sold fat to the butcher at fifteen months old. They are usually fed in the first instance in just the same way as the dairy heifer calves (described in Chapter iii) until they are three months old. At that age the heifer calves intended for the herd (and therefore not to be fattened) will begin to receive poorer foods, but the calves for baby beef must still have fattening rations. They are given good hay, good roots, and some mixed concentrated foods such as peas, beans, linseed cake, crushed oats and flaked maize. If this fattening time comes in the summer, when no roots are available, then it is usual to give them some green foods such as lucerne, oats and vetches, seeds mixtures, cabbages, or kale. No matter how good the weather may be, the calves are never allowed out on pasture, because they would then take too much exercise and would not fatten quickly. These same foods are continued after the calf is six months old, but at that age silage may be added to the other rations. Calves always like silage as soon as they are used to it, and always thrive well on it. Even as they get older they are not allowed out to graze, but are kept indoors entirely in yards and boxes. Usually about ten calves are kept together in a yard, or two to five together in a box. Calves do not seem to be happy unless they are in the company of others, and if they are kept alone in separate pens they will not thrive unless they can see and hear one another. If a calf or yearling is ill or lame it may be necessary to put it into a pen by itself, but sometimes it pines so much for company that it must be put back with its fellows. In some ways cattle seem to have very bad memories. If one beast from a yard is lame and is put away in a separate pen for treatment for a few days, when it returns the other beasts seem to have forgotten it and the lame animal may be badly bullied and treated as a stranger.

The fattening of cattle for baby beef is expensive because they are always housed. All the food they receive must therefore be taken to them, and much labour is used besides in supplying litter and in carting away the dung, for no machines can be of much help with these chores. Animals sold for baby beef must therefore be sold for a high price, or else the farmer would lose money by keeping them. They are sold young, when they cannot be very heavy. Thus farmers must receive more money per hundredweight of live weight for animals sold for baby beef than for beeflings or mature beef animals. In their turn those people who eat baby beef must pay the butcher a higher price per pound for it. Baby-beef meat is very tender, and it is lean; but on the other hand it is inclined to be watery, bony, and lacking in flavour.

Beeflings. Some of the animals kept for beeflings will be merely baby-beef animals that chanced to fatten too slowly, or that *grew* rather than *fattened*. These animals will therefore have been fed just as the baby-beef stock were fed, except that they will have been fed in this way for a longer period. During this extra time it may have been necessary to increase the amount of food each day, because these beeflings will by then be older and heavier than baby-beef stock.

The other beefling animals will be those that have fattened more quickly at a younger age than was originally anticipated. This group of beeflings will go through a store period, from six months to at least eighteen months of age. If it is summer they will be running out at grass, and will get nothing besides grass. If it is winter they will live principally upon inferior hay, straw, roots or silage, and practically no concentrated foods. It is rarely possible to *fatten* them on grass, and they are usually fattened in yards (p. 17) and boxes, being taken in from grazing in August as soon as the grass is affected by summer drought. If this is not possible, then as the grass disappears a little green food may be given on the pastures, and concentrates may be put out on the grass-land for them in tubs. But beeflings housed from the middle of August or early September can be fat in time for the Christmas market in November and early December. During this time they are fed on good hay, roots or silage, and some concentrates. The time taken to fatten them will be from four to six months. These beeflings are usually the offspring of dairy cows mated with beef bulls.

Mature beef.   The stock in this group can be fattened for market on grassland or in yards. The number being fattened in yards is declining annually because of the labour and expense involved, but the number fattened on grass has been maintained. Formerly store cattle were bought (by the arable-land farmers) in September, not always because fattening cattle was a main part of their business, but because the cattle were wanted to tread large quantities of straw into farmyard manure for use on the land in due course. They also ate what were really by-products of arable farms—such foods as roots, hay, second-rate cereals and pulses. Bullocks bought in the autumn in this way were fattened in four, five or six months. They were kept just long enough to consume the by-products that could not be sold. If there were not enough of these foods available to fatten the stock then farmers would have to sell off the cattle unprofitably as stores.

In the early part of this century, when bullocks were fattened they were fed on liberal rations of roots (up to

1½ cwt. per beast each day), and generous quantities of linseed and cotton cakes (4 lb. of each kind daily). But scientists have shown since then that this was a wasteful system (too much *protein* was being fed) and they advised giving more cereals (*carbohydrates*). Thus, at the present time (prior to the war in 1939) linseed and cotton cakes are but little used (only up to 2 lb. per head per day) and cereals are used freely instead (up to 8 lb. per head daily). Nowadays, in the early stages of fattening, only moderate quantities of concentrates (4 or 6 lb. daily) are given, but as the animals get fatter the quantity is gradually increased. The hay and roots too are usually slightly increased throughout the fattening period, but the percentage increases will not be as great as with the concentrated foods. Many farmers give their bullocks a little linseed cake throughout the last month of fattening. This makes their coats glossy, so that they look at their best in the sale ring. Silage, too, if fed throughout the fattening period, helps to produce a glossy coat.

In normal peace time appearance was important, because it helped the sale, but since farmers began to receive a special subsidy for the production of beef, appearance mattered less and *fatness* was of the first importance.

For centuries in England beef cattle have been fattened on grassland during the summer, but to fatten cattle during the winter is relatively new. Till two hundred years ago or so as many cattle as possible were killed off in the autumn, because it was difficult to find food for them during the winter. Fattening in winter, in fact, was not possible, but stock were fattened on grassland in summer to provide the winter's meat supply.

Cattle are fattened on natural grassland in the Midlands, e.g. Northamptonshire, Leicestershire, and the neighbouring counties, and on the marshes of Somerset, Sussex, and Lincolnshire. It is also possible to fatten cattle in any district where land has been sown with seeds leys for a spell of three or four years. The best grassland of all these kinds will fatten cattle without the need of any other foods. Sometimes the pastures are good enough to fatten cattle in the early part of the grazing season (May and June), but in times of drought the grass may be too poor for fattening purposes in August and September, and in some cases even in July. In other districts no pastures will be good enough to fatten cattle unless they get some concentrated foods in addition.

The cattle that are fattened on pastures are usually bought as stores either in the spring just when the grass begins to grow or in the previous autumn. Some farmers prefer to buy in the autumn because they want to be sure that the cattle live outside during the winter, since store cattle that are kept outside all winter fatten more readily on the grass in the following spring than do cattle that have wintered inside.

The cattle must be two or more years old when fattening commences, because younger animals (as we have seen) do not get fat on grassland. There are certain breeds that fatten most readily on grassland, e.g. Herefords, Welsh, Devons, South Devons, Sussex, Shorthorns (especially Lincoln Red Shorthorns), Galloways, and Highlands (whose horns are so large they could not be kept in buildings anyway). Crosses in which any of these breeds occur are very suitable for fattening on grassland.

When the grass begins to grow in the spring, stock should be matched up carefully for fattening on the grass; that is to say, the animals are sorted out into groups of the same sex and also the same degree of fatness. In the spring a group of cattle will be turned into a field, and there they will remain until they are fat. One acre of fattening pasture is usually allowed for every beast.

If the grass grows freely it is impossible to put more beasts into the field because of the fighting that would certainly result: but sheep can be added for a time, because sheep and cattle will not attack one another. When the supply of grass begins to fall off the sheep will be taken away again.

After three months of grazing some of the cattle will already be fat enough to be sold, and if the farmer has matched his stock well, then all that are in any one field will get fat within the space of about a month, by about the end of July. Since the grass should continue growing into the autumn, a second group of stock may now be turned out to fatten in the same field, but in this case either a smaller number will be put into the field or concentrated foods will be given in addition to the grass they eat. This second group may be fat by September or October. Some that are nearly fat at the end of the grazing season are put into yards and fattened as arable-land stock in late autumn and early winter. It is essential to graze the fattening pastures thoroughly before the winter, and some store cattle may be used for this final grazing, and perhaps will be bought for this purpose in the autumn so as to be ready for fattening on the pasture in the following spring. All this applies to the best pastures and leys only. Concentrated foods such as cakes and cereals may be fed to cattle on second-rate pastures throughout the whole of the fattening period.

Cow beef. Cows culled from beef and dual-purpose herds may be fattened in various ways, namely in yards in winter, on pasture in summer, or while still in dairy herds. In yards and on grass the fattening will be just the same as has been described for the mature fat cattle; but the fattening in the dairy herds calls for some further explanation. If cows are found to be unsuitable for milk production—for example, if they are ceasing to breed or have udder troubles—they may be fattened while still in milk. This is done by giving them more concentrates than is their due according to the milk they are producing. Sometimes it is possible to fatten a cow so easily in this way that she may be fat when she goes dry at the end of her lactation. But generally a cow is dry for a month or two before she is fat enough to be sold. In this case the dry cows are fattened indoors and rarely on pastures. Fat cows usually sell best in the winter, because there is a much greater demand in winter months for the beef they provide for boiling, and little demand for it in the summer. The price fetched is usually highest of all in November, because fat cows produce a large quantity of suet, and suet is needed in November for the making of Christmas puddings.

CHAPTER V. SHEEP

THERE are three groups of sheep in Britain, each group named after the kind of land on which they are kept, namely, mountain sheep, grassland sheep and arable-land sheep. The feeding and treatment of sheep depends mainly upon which of these groups they belong to. Within each group there are a number of different breeds and crosses. In fact, there are more different breeds of sheep in this country than there are of any other kinds of live stock, because sheep are so much affected by local conditions. Then, farmers often make crosses between rams of one breed and ewes of another so that among the various sheep sold at local markets it is often difficult to say to what breed they belong. As the treatment for each of the three groups is very different from that of the others it will be best to describe each group separately.

Mountain sheep are small, active animals. The food supply on mountain pastures is so limited that only very small, active sheep could hunt sufficiently for it (p. 17). Breeding takes place in the mountains, but fattening of the sheep there is impossible. In parts of Wales, where the farm boundaries are not defined by fences, hedges or walls, mountain sheep are exceptionally valuable because of their knowledge of the boundaries of the farm. All that graze there are born there and the breeding stock remain there. In these cases, fresh sheep are never introduced to the Hill. The rams of one farm will drive off any sheep that may stray from other flocks into their mountain grazing.

The mountain sheep are so hardy that they are able to live and grow and find food enough where other breeds would starve. Sometimes the mountain sheep are so used to finding grass for themselves they will not eat hay, or concentrated foods even if they are provided. And experienced farmers are against giving these foods to the sheep anyhow, because such feeding spoils them, so that they would not then forage around sufficiently to find their own livings. Therefore, since the pastures afford little enough food and nothing extra is given, a farmer must plan very carefully when the lambs shall be born. He must be sure there will be pasture enough for the ewes when they are producing milk; it is then that they need most food. So with mountain sheep the rams are turned amongst the ewes on the mountains at the end of October or in early November, and lambs are therefore commonly born at the end of April and in early May, when the ewes are on the lower slopes of the mountains. They are there because the upper slopes may still be covered with snow. The flocks are scattered over a wide area, and the shepherd cannot take as much interest in the birth of his mountain lambs as do the shepherds who care for flocks of arable-land sheep. Mountain shepherds, too, believe that they must not coddle the ewes and lambs since that might save the weaker ones and end by producing sheep that could not stand the cold and exposure of the mountain climate. Sometimes, because of late falls of snow, or heavy rains bringing mountain torrents, many lambs may die. Although it may seem callous and unkind, the farmers say that such rough treatment leads to the survival of the fittest.

As soon as all the lambs have been born (i.e. by the end of May) the ewes and lambs are sent up to the mountain side. The shepherd is always about on the mountain side with them; his objects are to encourage the sheep to graze where he wishes them to, and to see that ewes and lambs are not in difficulties. Sometimes ewes and lambs will get into gullies or ravines or on awkward ledges, or they may be caught up in bushes, and the shepherd must be at hand ready to rescue them. As the summer advances, so the ewes and lambs graze the higher slopes. In June, July, or August the ewes are shorn, and as far as possible they will remain up the mountain side for this. By law, after shearing time and between certain dates the sheep must all be 'dipped' (p. 32), that is, the sheep must be dropped into a tank containing a solution of certain chemicals with which they are thus wetted from head to foot. Dipping is done to kill the parasite that causes scab, and to kill ticks and lice. In some districts the law decrees that all the sheep must be dipped twice within a stated period.

By the end of August the lambs are 'weaned'. This is done by bringing the lambs to the lower slopes and leaving the ewes on the mountains to graze the highest parts. After about a week the ewes will cease to produce milk and the lambs may be put back with them. The whole flock will be set to graze as high up the mountains as possible, so that the food on the lower slopes may be reserved for winter feeding.

The food on the mountain is so poor that it is impossible to fatten lambs on the mountain slopes (p. 17), and it is necessary to sell them to be fattened elsewhere. The sales of the lambs, which are called store lambs, take place in the autumn. Store lambs are usually bought by farmers who have grown roots especially for fattening them. Sometimes the farms to which the lambs are sold are situated in the valleys of the mountains; but often they are far away and the lambs may have to be moved long distances from the place of their birth. Other mountain lambs are sold to be fattened on good grassland; but this fattening will not begin until the following spring when the new growth of grass occurs. The lambs sold will always be wethers (males castrated), and only rarely any ewe lambs, as these will be wanted to maintain breeding flocks under mountain conditions. Mostly, all the ewe lambs born are kept for the flocks. Each year a certain number of older ewes are culled and sold. These ewes are usually culled because they are lame, and since they are quite suitable for breeding they are sold for breeding under less rigorous conditions than on the mountain side. These ewes form what are known as 'flying flocks', which will be described in greater detail later. Of mountain sheep farming it should be emphasized that food and climatic conditions are usually so bad that no sheep at all will fatten there—not even ewes; so they too must be sold to fatten elsewhere.

Grassland sheep. As the mountain sheep include the smallest breeds of sheep, so the grassland sheep include the largest breeds in the country. The quality of the wool produced is usually much better than that from mountain sheep, and the quantity is often twice, three times or even four times as much as that from the mountain breeds. The meat is quite different too; mountain sheep produce small lean joints of tasty meat, whereas the grassland breeds, in general, produce large joints of fat meat, rather lacking in flavour. The grassland breeds have been widely exported to various parts of the world where stock were required to feed on grassland or on 'range' as it is usually called abroad. In Great Britain the grassland sheep are found in many districts, but more especially in the Midlands. The Romney Marsh or Kent Breed deserve special mention. They have lived for years as the only occupants of Romney Marsh, and have been widely exported. The Kent breed has played a big part in building up the New Zealand Canterbury Lamb trade. In recent years, though the numbers of the grassland breeds have been declining, the number of sheep actually living on grassland has not, because mountain-bred sheep are now being bought for breeding on grassland (p. 80). The reason for this change is that joints from the mountain breeds are of a size best liked by the public, whereas joints cut from the real grassland breeds are too big. The breeding and rearing of grassland sheep will be the same whatever the actual breed.

Grassland sheep have their lambs either just before or just after the grass begins to grow in the spring; if just before, then their arrival is carefully planned so that the grass will be ready for the lambs when they are old enough to eat grass. But with this early lambing it is essential to have a supply of hay and roots at hand to feed to the ewes until the grass grows, and while they are producing milk for their lambs. In districts where there is nothing but grass, the farmers will arrange for the ewes to have their lambs later, when the grass has grown, since grass is the only food available. In short, breeding must be carefully timed in the autumn so that the ewes produce the lambs at the correct time, according to the district and food supplies.

Just before the rams are turned into the breeding flocks the ewes are 'flushed', that is to say, they are given particularly good food for two or three weeks. This results in more of them being ready for breeding, and also in more of them having twins. Secondly, some farmers divide their flocks into groups of about sixty ewes, putting one ram to each group to remain with the group for fourteen to twenty-one days, another ram being substituted at the end of that period. Rams will remain with the ewes for about two months in all, though most of the ewes should breed within a month after the rams are first added.

After the rams have been with them, the ewes may remain grazing on grassland, seeds leys, or even sugar-beet tops or roots, if these are available. Often nothing

4-2

will be given in addition to these foods, except to those ewes that look a little thin. The thin ewes may be put into a special group to receive extra food. They may be the old ewes that have shed some teeth, or young ones that have not grown all their teeth, or ewes that are carrying twins or triplets. The special food may consist of chaffed hay, chopped roots (because of the teeth trouble) and concentrated foods such as crushed oats, flaked maize, cracked peas, and linseed cake. As the time of lambing approaches, further ewes may be added to those receiving special rations.

For the best results special care of the ewes is needed at lambing time. Some farmers have no shepherd, and instruct anyone who has any spare time to look after the sheep. With unskilled attention the sheep are likely to be neglected, and an unsatisfactory fall (or survival) of lambs may be expected.

Farmers who take special interest in their sheep will set aside a number of grass fields for the ewes at lambing time. In one field the ewes that have not yet had their lambs will be kept. They will remain there until the lambs are born, and will then be moved into a second field, where they are kept in small pens (a ewe and her lambs to a pen) for twenty-four hours. This is done to make sure that mother and offspring get on well together, and to ensure that the lambs are properly fed. After twenty-four hours the ewes and the lambs are turned loose into this second field where they remain for several days under observation. If all goes well with them they are moved to field 3 or 4, according to the policy of the farmer. He may decide to divide the ewes into two groups: those with single lambs into field 3, and those with twins into field 4. Or he may group them according to the age of these lambs instead of according to the number of lambs per ewe. If he follows this method fifty or sixty ewes and their lambs will be placed in field 3; the next fifty to bear lambs will then be collected in field 4, and in further fields. This making of small flocks will continue until all the lambs are born. All lambs are tailed and most of the ram lambs castrated in the first month.

How these ewes and lambs are treated when lambing is completed will depend entirely upon the aim of the farmer. He may plan to sell the lambs fat, or to sell all the lambs as store lambs, or to sell some fat and some as stores. If the lambs are to be sold fat, he may let them fatten as they will on the grass that is available, or in times of plenty he may give them some concentrated foods in addition. Sometimes the group of ewes that have twin lambs may have the better food while the ewes with single lambs may be left to fend for themselves. Lambs that are to be sold as stores (and the ewe lambs that are to join the farmer's own breeding flocks) will receive no special foods—they will merely eat grass; and the ewe lambs for the flock will live as stores in this way until breeding commences. Lambs that are sold as stores for fattening may go to arable-land districts to be fattened on root crops during the winter, or they may go to the grassland fattening districts where they will be fattened along with the cattle, as mentioned on p. 45. They are there eventually to be fattened on grassland; they will live on the grass during winter as stores in their new homes, and get fat on the spring grass.

After their lambs have been taken away, the grassland ewes are used as scavengers on the farm, eating any surplus grass or any second growth of seeds from the arable land; or they may be folded on roots not wanted for other purposes. Thus for a time some grassland sheep may live on arable land; though it still remains true of the grassland group as a whole that they generally live on nothing but grass.

The fattening on good grassland is a very simple business; the sheep are merely moved to the grass fields as
necessary to keep the grass just the right length for the
fattening cattle. These sheep do not usually receive any
concentrated foods in addition to the grass.

Arable-land sheep. The most important single fact
to remember about arable-land sheep is that they grow
and fatten quickly. They are bred for early fat-lamb
production and also are crossed with other breeds to
produce quick-growing lambs for the butcher. Arable-
land sheep are moderate in size and produce quite a
good quality of wool. They have not been widely exported,
and the number kept in the British Isles has declined
rapidly because to keep them costs so much care and
work. They are often kept entirely on the arable land,
which means that a full-time shepherd will be wanted
to look after 250–300 breeding ewes and their lambs.
As they are kept on the arable land in folds made of
hurdles or wire netting, the folds must be moved every
day to provide fresh food for the sheep, and also to ensure
that the manure is spread well over the field. Troughs
and hay racks must also be moved each day, or else the
sheep will have too much dung around those places
where they stand continually and the next crop planted
or drilled on the land will grow unevenly, more richly
at the well-dunged places. Folding calls for skill, for the
shepherd must either see that the field is evenly manured,
or, if the soil on the field is variable, then he must avoid
heavy manuring on the best soil but be sure that heavy
manuring goes to the poorer places by putting the troughs
there. He must allow more fold space in wet weather
because the sheep will puddle the surface, and may spoil
the texture of the soil if they are crowded together then.
With arable-land sheep, too (and this is important), care

must be taken in moving them from one field to another to see that the changes do not upset the sheep. Sudden changes of food may make sheep ill.

Most of the arable-land ewes have their lambs at the beginning of the year, but in Hampshire and the neighbouring counties one breed, the Dorset Horn ewes, may have their lambs at any time of the year, and many may be bred so that lambs are fat for Christmas; but as Dorset Horn sheep are not widely kept, we shall omit further mention of them and describe the management of general breeds only.

In the month of July the breeding ewes may be on seeds leys or even on grassland. About three weeks before the rams are turned in with them for breeding the ewes are flushed by putting them on good aftermath or seeds leys for a period of two or three weeks. Sometimes the ewes are grouped in sixties (as with the grassland sheep) for the breeding, one ram being assigned to each group; at other times all the ewes are kept together and two rams are put with the whole flock. With these sheep occasionally, as with the grassland sheep also, the rams are marked with colour on their chests so that they mark the backs of the ewes they mate with. By watching the markings the farmer and the shepherd can tell how many of the ewes are mating. The rams remain with the ewes for about two months, and during this time the ewes will be fed on seeds leys and possibly on turnips, kale and cabbage. In October sugar-beet tops are ready, and these will be fed throughout the remainder of the year. All arable-land ewes should be given hay at all times, and this is given in racks. Hay is supplied for various reasons, but especially to counteract the laxative effect of the root crops. Sheep do not generally drink water when they are folded on roots, but if they are folded on a crop of seeds or seeds aftermath they require water, especially if the weather is hot. If any of the ewes are in poor condition they should be put into a flock where they can receive concentrated foods.

Careful preparations must be made for the lambing, which usually comes in January and February. A special lambing pen (p. **32**) is made to provide a warm sheltered yard for the lambs. The lambing pen has three or four large compartments. In (*a*) are the ewes that have not lambed; in (*b*) the ewes that are lambing; in (*c*) the ewes that have single lambs; and in (*d*) the ewes that have twin lambs. If space is limited one compartment will take both (*a*) and (*b*) groups. Around the lambing yards are a large number of small pens each for a single ewe and her lamb or lambs. To protect it against wind the whole yard must be walled round with straw and hurdles. Some of the lambs with early lambing will be born in severe weather, and some may even die in spite of all such precautions. The shepherd has a special hut with a stove in it, and he remains with the flock all the time during the main lambing period, about a month. Many a shepherd will proudly tell you that 'He don't take his clothes off for a month' at lambing time.

As the lambs are born, each ewe and her lamb or twins are kept in a pen for a day or two to make sure that they are used to one another, and then the ewe will be moved into one of two larger pens, according to whether she has a single lamb or twins. On the first fine day, when the lambs are a week or two old, the shepherd moves the ewe and her lamb or lambs from the lambing yard to a fold. For several days the ewes and lambs will go out to the fold by day and return to the lambing yard at night. When the weather is not too severe these ewes and their lambs will stay out on the fold all night, but to afford extra shelter lamb sheets are hung on the hurdles; these are made of sacking and are as high as the hurdles and they keep off the worst of the wind.

At this time of the year the ewes and the lambs are folded upon swedes, turnips and kale, but the ewes also need hay and some concentrated foods. But, until they are two to three weeks old, the lambs will not take much food besides the milk from their mothers. To ensure that the lambs, as they get older, shall receive their fair share of food, they are often allowed to go alone into a fold one day before the ewes enter it. This is simply done by the use of a special hurdle, in which vertical rollers are fitted close enough to let the lambs get between the rollers, but not the ewes. The farmers call this way of giving the lambs the first bite of the best food 'letting the lambs forward in the creeps'. Special concentrated foods are also put out for the lambs in their own fold. Sometimes a ewe has three lambs, though she may have only milk enough for two; the third lamb is then taken away and, if possible, given to a foster mother who has no lambs or only one of her own lambs surviving. It sometimes happens, too, that a ewe dies and leaves some orphans; these also are either given to foster mothers or they are reared by hand. Shepherds do not like hand rearing because of the work it means; for young lambs must have three or four feeds of milk a day, and care must be taken each time to give the milk at body temperature, otherwise the lambs will suffer from stomach-ache and diarrhoea. Sometimes no foster mother is available and it is necessary to feed a lamb on the bottle till a ewe appears that needs another lamb; but there is a risk that once a lamb has learned to drink from a bottle it may not change to sucking a ewe. To obtain milk for orphan lambs some farmers who are not dairy farmers buy one or two freshly calved cows when lambing commences. Again, lambs running with ewes will have milk from their dams for three or four months, but the shepherd finds time to bottle-feed lambs for only about two months; and as on so short a period of milk lambs will not grow as fast as usual, bottle-reared lambs will not be fattened as lambs, but as mature sheep.

Most of the arable-land lambs are fattened as quickly as possible. They are given plenty of concentrated foods and a good supply of roots, and some will do so well that they can be sold fat at three months old, and others— the majority—at four months old. All these lambs will be getting milk from the ewes till the day they are sold fat. On some farms, however, and with some breeds, the lambs are not fattened quickly but are kept as stores until they are weaned; if this is the plan, then they will not be given any concentrated foods.

Store lambs from mountain, grassland and arable-land flocks may be fattened on the arable land from July onwards according to food supplies. In July and August the principal food is aftermath seeds. From September onwards roots are often available. Some few farmers breed their own sheep as stores to be fattened upon these foods, but most farmers buy the young store sheep, the number they buy depending upon what food the farm can supply. These sheep will take from three to six months to get fat, and will be sold fat between the ages of eight to fifteen months. They will be fed on hay, some concentrated foods, and an abundant supply of roots. When roots are fed no water is needed; but if the sheep are on seeds leys water is essential for quick fattening.

Old ewes which are not to be used again for breeding are sometimes fattened off while suckling, but more often they are kept till the following winter and then fattened on roots. To help those ewes which have very few teeth the hay is often chopped or chaffed and the roots are sliced or pulped.

Flying flocks. Some farmers buy breeding ewes from mountain, grassland, and arable-land flocks and breed from them for just one season. After the lambs are born the ewes and their lambs are fattened at the same time by generous feeding. This is most commonly done on arable land, but it may also be done on grassland. The lambs are sold at about four months old, and the ewes remain on the farm for no longer than ten months in all. Because of this short stay on a farm they are known as flying flocks. Ewes of various breeds are included, and are mated with rams of the local breeds.

Wool. Though this chapter deals with the raising of sheep and lambs for the meat market, it would hardly be complete unless mention were made here of wool. Sheep in Britain are shorn between May and August (the date depending on the district, the weather in a particular year, and the breed of sheep). A fleece is a warm and heavy covering and it is not removed until the weather is so warm that the sheep will not catch cold without its protection. At the beginning of this century it was the custom to wash all sheep before shearing, but manufacturers now prefer to wash the wool themselves at the factories, so that nowadays washing does not come before shearing. Formerly the shearing was done entirely by hand shears (p. 33), but shearing machines, driven either by an engine or by electricity, are now in general use.

After the wool has been shorn, each fleece is carefully rolled up, tied with wool, and specially packed for the manufacturers. Pieces of hay, straw, dried grass and string must be kept out of the wool, and the fleeces packed into *jute* bags before despatch. These packing and cleaning precautions are necessary because all foreign fibres are a nuisance in the manufacturing process, they cannot afterwards be separated from the wool, they are brittle, and do not take the dye properly.

All sheep that are over one year old are shorn in May, June or July, the lambs being omitted in this country, though in some other countries they may also be shorn. The quantity of wool obtained from a sheep depends on the following points:

(a) Closeness of clipping (mountain sheep are not closely clipped).

(b) Breed (the bigger breeds usually produce more wool).

(c) Sex (rams usually produce more wool than ewes).

(d) The life of the ewes (ewes that have had lambs in a season give less wool than those that have not).

(e) Nutrition (well-nourished sheep give more wool than the poorly fed).

The actual weights of fleece vary a great deal and range from 3 lb. to 16 lb. per fleece. Not only does quantity of wool vary but also quality: wool from mountain sheep is often so coarse that it can only be used for carpet-making; from the other breeds the quality varies with the breed; some of this is used for suitings, and the finest for underclothes. But nowadays the principal product of British flocks is not wool but meat.

CHAPTER VI. PIGS

THERE are now no wild pigs in Britain; but we know that, when there were, the sows 'farrowed' (had young) once a year only, and their litters were small, from two to four pigs to the litter. Wild pigs lived for the most part in woods, and rootled around with their snouts for such food as they could find. Food was often scarce, and the wild pigs grew slowly. Many were killed each autumn when they were several years old. At first when pigs were domesticated they were housed in very bad conditions—perhaps those who kept them thought that as pigs were hardy by nature they should not be coddled up in warm buildings—and until quite recently little thought was given to their care and housing.

During the last twenty years, however, great changes have taken place. The better-kept modern sows will rear a litter of eight to twelve piglings twice every year. The piglings will be fed principally upon miller's offals and ground cereals, and under this treatment will grow so quickly on peace-time rations that they are fit to kill for pork or bacon at six months old. There has also been a decided movement to keep pigs in clean, warm, dry buildings; and results prove again and again that pigs thrive for fattening much better so. Nowadays, also, it has become the practice to keep breeding pigs out of doors, not in woods as formerly but on pastures (p. 33) or on land on which suitable crops have been grown; and undoubtedly the outdoor system leads to healthier breeding stock, and the young pigs also thrive better if kept properly out of doors than under the old bad conditions indoors. From district to district, however, big differences are still found in the treatment of pigs.

Indoor systems for sows and boars. At the present time, breeding sows and boars may be kept either
- (*a*) outside; on pastures (p. **33**), or in woods, or in runs; or
- (*b*) inside; in straw yards (p. **64**); or
- (*c*) inside; in boxes or pens (p. **64**) or sties.

(*a*) On pastures the sows live naturally, taking as much exercise as they like, eating green foods, and receiving the benefit of basking in the sun. They should not remain on the same land for more than six months because the land may become infested with internal parasites. Wherever sows are kept out of doors great care is needed to prevent them from breaking down fences and straying away. The muscles of a sow's neck are very strong; if she puts her snout under a fence or a gate she may easily lift it and walk out. A strand of barbed wire fixed along a fence about six inches above ground will prevent a sow from trying to break through. Electric wire fencing has also been used with excellent results to keep sows within bounds. Sows kept out of doors will need a certain allowance of solid food, but their access to natural green foods will reduce the amount of meal required.

(*b*) Sows always seem to be very comfortable in yards if they are provided with plenty of straw during cold weather. In the summer much less straw is needed; and in very hot weather sows greatly enjoy a mud bath. They will thrive best if the yard is so arranged that part of it is always in shade and part in sun, so that they can choose where to lie according to their fancy of the moment.

(*c*) Many sows, however, are still kept in the old-fashioned boxes, pens, or sties; though in these places they get very little room for exercise, and often no direct sunlight. All breeding sows, when kept indoors, must be given some meal; and they will thrive best if they receive some green stuff as well.

Wherever she has been kept, a breeding sow, about ten days[1] before her piglings are due to be born, must be moved to the place in which she is to have them (her farrowing quarters), so that she will be quite settled down in the new surroundings when the piglings are born.

For many years sows have been moved to boxes or sties for this purpose. The only merits of such a system are that a large number of pigs may be kept in a small area, and that, as the sties are generally near at hand help may be given without trouble if a sow should need it at farrowing time. In a Danish piggery (p. 64) each sow will have a pen under cover for herself and her piglings. But if piglings remain indoors completely they sometimes suffer from anaemia after they are three weeks old; the better Danish piggeries have outside runs, one for each pen, so made that sows and piglings can go in and out at will.

When the time comes to move them out of boxes, sties, or Danish piggery where the litter was born, several sows and their litters may be turned together into a yard (either partly or completely covered); or the piglings may be moved permanently to outdoor runs as soon as they are two weeks old, or three weeks, if it is winter time.

Outdoor systems for sows. Of the outdoor systems, a shed with a run is the oldest, but there is little good to be said for it. Such runs are usually so small that there is very little grass on them; as the sheds are difficult to move, the same area of ground is likely to be used for many months and even for years, so that the ground of the run will be infested with the eggs of internal parasites, the piglings will suffer from worms, and will be very unthrifty in spite of the fact that they are being kept out of doors.

---

[1] Fourteen days for a gilt.

During the last few years various forms of folding have been developed to make it easier for the ground to be changed far more often. Huts with very small pens attached have been designed so that hut and pen together can and should be moved to a fresh site daily (p. **33**). These may be quite detached (each hut and pen a unit) or several such huts and pens may adjoin one another, in which case less fencing materials are needed. Sows and litters kept in these folding units thrive very well; but the system is not popular because it is expensive: the portable sheds and pens cost a great deal to buy in the first place, and the daily labour required to move the pens is a big item of expense.

Another outdoor method, the tethering system, does not cost nearly so much for labour, because at some times of the year the housing unit may not need to be moved more than once in a month. With this system a *small* hut is provided which is just big enough for the sow to lie down lengthwise in the hut but not across it. The sow has special harness fitted round her neck and shoulders, and is attached by this to a chain 10 to 12 yards long, fixed to an iron peg driven into the ground. This peg is so placed that the length of chain will allow the sow to get in and out of her hut, but not to go round it and get the chain entangled. The piglings are allowed to roam but they return for suckling. The tethering system is good and leads to the production of very sturdy piglings, but some people object that the piglings roam one and even two fields away from the field where their huts are. Thus in herds of pedigree stock, where piglings must be periodically weighed, a pigling may stray from its mother to a sow that is not its mother and feed from her; if this happens repeatedly there may be great differences of growth between the piglings of one litter, and as a pedigree sow's efficiency is measured by the weight

9. Fattening pigs in a straw yard

10. Fattening pigs in a Danish piggery in England

11. Mares and foals at pasture

12. Breaking in with reins

of her piglings, if any of them have been poaching quite a wrong opinion may be formed of their mother's quality.

Piglings: birth and rearing. Whatever system of housing is followed, it is essential to give every sow time enough to get properly settled down in her new surroundings before her young are born. If not, she may be so nervous that she will attack and eat her piglings. Some sows that are usually quite docile may be very savage for a few days after their piglings are born, and it is unwise to disturb *any* sow more than is absolutely necessary for several days after she has farrowed. The sows are fed lightly for the first twenty-four hours after farrowing, after which they must have plenty of food because they produce very rich milk, so rich, in fact, that piglings grow faster than the young of any other farm animals: they weigh about 3 lb. when born, and by the time they are eight weeks old they may be ten or occasionally twenty times their birth weights.

At three weeks old they begin to eat solid food; and wise farmers at once supply special rations for these pigs. Weatings, both wet and dry, are very popular food for piglings; whole wheat, too, is good, it is sometimes given mixed with peas. Some arrangement must be made so that food may be at all times within reach of the piglings, but in places where the sow cannot get at it, otherwise none may be left for the piglings. Special gaps in fences may be made for the piglings, or pens may be partly partitioned off so that the piglings can enter the partition but the sow cannot. At six weeks old, any boar (or male) pigs not to be kept for breeding are castrated. An age of six weeks is chosen so that they may have two further weeks before they are weaned in which to recover from the operation.

During the last few years a method of judging the value of sows by weighing their progeny has been increasingly followed; this is of course an indirect way of measuring the milk production of the sow. For this purpose the piglings may be weighed, either

(a) at three weeks, up to which time they have had nothing besides sow's milk; or

(b) at six weeks, before castration; by this time they have had some food in addition to sow's milk; or

(c) at eight weeks, when weaning is about to take place.

Each of these alternatives is open to objections:

(a) If the piglings are weighed as young as three weeks old there is no record of the milk produced subsequently by the sow, though that is most important in judging her quality.

(b) Even if they are weighed at six weeks the full record of milk produced is not obtained; also at this time the sow may get the credit of an increase in weight really due to supplementary foods given to her piglings.

(c) If they are weighed at eight weeks the sow will be credited with all the milk she has produced but the weight of the piglings may then be largely due to the supplementary foods they have received.

Weaning and afterwards. At weaning time the sow is separated from her piglings and in about three days' time she is ready to go to the boar, and so the breeding cycle continues. At weaning time only the boar piglings that will later on be required for breeding are removed from the litters. The rest of the weaned piglings are handled in various ways; the whole litter may be placed in a box or pen; or several litters may be put together into a yard—sometimes as many as a hundred piglings may thus be kept together; or, if the piglings have already been living out of doors, litters may be grouped together and left on pastures, such a group may also consist of fifty to a hundred piglings.

After weaning, piglings are generally fed on meal mixtures, such as weatings and fish meal; or weatings, barley meal, flaked maize, meat meal and minerals. The minerals added to the food are chalk and phosphates and a trace of salt; salt in quantity is almost a poison to pigs. Water should always be at hand for pigs to drink whenever they need it; newly weaned piglings may be particularly thirsty.

Until they are at least three or four months old, the gilts (young sows for breeding) and the pigs intended for fattening remain together. Some farmers do not decide which to keep as breeding gilts until all have reached the weight required for meat production; they find they can judge the shape of a pig better when it is fully developed for meat than at any younger age. Others believe that it is better to select the gilts when they are about three months old, and to keep them on store rations (i.e. growing but not fattening rations) until they are actually having their young, which will be when they are twelve to eighteen months old. By this second method all gilts of approximately the same age are kept on a pasture or in a straw yard, and given 'growing' rations until they are big enough for breeding. They thus take plenty of exercise and become strong and healthy, which is what is wanted in the breeding stock.

The other system of rearing—fattening *all* the group and selecting the breeding gilts when they are of market size—will be described in detail later; here, it is enough to note that the fattened gilts that are to be used for breeding must first of all be slimmed down, and that this must be done with care, without upsetting them or running the risk of chills. As, while they were fattening, they were housed and kept warm, the change over to stores must be gradual, and they may have to remain indoors for some further time until the weather is just right to turn them out.

Breeding. When gilts are eight to twelve months old —the exact age depends upon breed, development and convenience—breeding may commence. At breeding time, with pedigree herds, the gilts are kept near a boar and mated as they are ready. Breeding will thus be controlled and recorded. But in many commercial herds a boar may be turned in with a group of ten to twenty gilts or sows to breed without record. No change at all is made in the feeding and management of the gilt when she is in pig until two weeks[1] before the piglings are due to be born, when she must be moved into the farrowing quarters in good time, so that she may settle down thoroughly before their arrival. (Occasionally, a gilt may be thin and demand extra feeding during the latter half of the time before her piglings are born, but this is rather unusual.) A gilt does not become a 'sow' with her first litter of pigs, but with the second litter; by that time she is from eighteen months to twenty-four months old.

Classes of pigs for the meat market. In normal times several different kinds of pig-meat are produced. For the London pork trade pigs will be killed at the age of four to five months, weighing 90 lb. alive or 60 lb. as a dressed carcase. Two British breeds—the Middle Whites and the Berkshires—are most suited to the production of this kind of meat. Both of these breeds are short and rather fat; they grow quickly and fatten readily while still young.

[1] Ten days for a sow.

For the Provincial pork trade a bigger pig is wanted, weighing 140 lb. alive and nearly 100 lb. when dead; these pigs are killed when five months old.

The next groups are those for Wiltshire Bacon which are ready at six months, when they should weigh 200–250 lb. live weight, or 140–180 lb. dead weight. For the Wiltshire trade long and fairly lean pigs are needed. In Denmark, famous for its bacon, pigs from the English breed known as the Large White are reckoned the most satisfactory for Danish Bacon. In England the best bacon pigs are obtained by cross-breeding Large White boars with Large Black Essex or Wessex Saddleback sows. These cross-breeds are known as Blue Pigs because of their colour. Blue pigs have what is known as hybrid vigour, i.e. the progeny of the cross are better than either parent; they fatten very rapidly, faster than a pure-bred pig of any one of the parent breeds.

For the Midlands and North of England, pork pigs eight or nine months old are killed when they weigh 300 lb. alive or 225 lb. when dead. These pigs are fairly fat and several breeds provide the kind of pig that suits this trade. It is worth noting that young pigs as killed for the London trade, weighing 90 lb. alive, will weigh only two-thirds of that weight, 60 lb., as dressed carcases; it is a general rule that the older the pig at killing the less in proportion is the difference between live weight and dead weight.

There is still one other kind of pig-meat produced and that is used in making sausages and pork pies, namely that from sows and boars. These are killed when their breeding days are over; their weights are more variable, and some breeds of pigs may weigh as much as half a ton.

Fattening. The actual methods of fattening depend on food supplies, on the accommodation available, and on the age at which the pigs are to be ready for market. In New Zealand up to 20 gallons of whey are fed to a fattening pig each day. In Denmark skimmed milk and whey are both widely used for pig feeding. In parts of the West of England, also, dairy products are commonly given to pigs, but there the quantities given are very much lower than in Denmark and New Zealand, simply because large supplies of such food are not available. It is well known, however, that pigs thrive well upon dairy products; the quality of meat produced also is first class; hence the phrase 'Dairy-fed Pork' which in peace time is sometimes to be seen in butchers' shops. Dairy products are found to be most effective when given to pigs of three or four months old; they therefore form an important part of the food given to pigs for pork (of 90 and 140 lb. live weight) or for bacon.

Potatoes are widely used for pig feeding, and it is quite common to find pigs being fattened in potato-growing districts; in any crop there will always be small or broken potatoes unfit for the market, which can most easily be turned to profit by keeping pigs to eat them. To make them more digestible potatoes should always be cooked for pigs.

In many places swill (or food waste) is available from hotels, colleges, schools and institutions. In war time, in many towns, there has been a house-to-house collection of household food scraps. This consists of vegetable trimmings, scraps and scrapings from plates and, in peace time, of meat scraps and bread. The law insists that swill that has been in contact with uncooked meat must be boiled before it is fed to pigs. Such swill can be fed to all pigs over three months old: younger pigs will not digest it well. It is true that bacon factories at one time used to complain that the carcases from pigs fed on swill were not very good for bacon production, but in war time such criticism seems to have disappeared. If swill is fed then, a basic ration in addition to about 2 lb. of mixed meal should be given each day. As a war-time measure many pigs have been fed on mangolds, kale, grass, and even on silage. A saving in the amount of meal needed can thus be made, but large quantities of these foods should never be given, or the pigs will not fatten or will fatten too slowly. Ordinary grass silage is *not* very suitable for pig feeding, and in general better results will be obtained if this food is given to other farm animals. The best silage for pig feeding is made from sugar-beet tops or potatoes.

In normal times many pigs were fed upon meals alone and were given none of the various foods mentioned above. A few years ago fish meal was first fed to pigs with excellent results, but it was soon found that certain kinds of fish meal on the market contained too high a proportion of oil, and the resulting bacon was found to have a fishy taint. Small quantities of fish meal were found to produce such excellent results that some farmers were tempted to increase the amount they fed daily; this also led to tainted bacon. It is, however, quite safe to feed *good quality* fish meal in *small quantities* for bacon production without the least risk of a taint. The most common cereal meals used for fattening pigs are barley, maize, wheat and weatings.

Many pigs are fattened intensely for only three or four months; others are fattened continuously from weaning time. Thus some pigs will enter their fattening pens at weaning; but where it is possible pigs will fatten best if they are not immediately moved into fresh surroundings at weaning, but at some later age, at three months old for example. The fattening pens may be in the form of a Danish piggery, a litter of eight or ten pigs to a pen (p. **64**); or may be boxes or sties. According to another system fifty to a hundred pigs are kept in a covered yard together, generously supplied with straw. On some farms the pigs complete their fattening altogether in these yards: but on others they are moved into boxes for the last month of fattening. It is a good plan to keep the pigs at weaning time in big groups together and later to match up eight or ten of equal size to fill a fattening pen; it is quite likely the whole pen will thus be fat together, whereas if pigs of one litter are kept together, some of the smaller pigs will almost certainly be slower in fattening than the bigger ones.

Sows are fattened in boxes or yards. If they have had only small litters the fattening may begin while they are still suckling. They will not often, however, become fully fat while still suckling their pigs, and after weaning a further two months may very well be needed.

For the 'small pork' trade the pigs are fattened as rapidly as possible on nothing but meals and dairy products. The medium-sized pork pigs are handled similarly, except that they may do without the milk products. Bacon pigs are fed on meals, with or without potatoes or swill, and in general are not given milk products at all. Formerly, bacon pigs were given as much food as they could eat, but since lean bacon is now popular it has been found that to obtain the best results feeding should be slightly restricted. The large pork pigs and the sows are fed in the same way as the bacon pigs, but the quantities of foods given are much greater.

It will be seen that pigs, in regard to the foods they eat, are similar to poultry but very different from all other farm stock; they need plenty of meal, they do not eat hay or straw at all, and only small quantities of roots.

CHAPTER VII. FROM THE FOAL TO THE WORKING HORSE

WE have already seen, in Chapter I, that in spite of the increasing use of tractors, especially on the biggest farms, there is likely to be a continuing place for horses also. The larger farms, where tractors are kept, will always need horses for jobs such as carting of food and litter to animals kept in buildings. The smaller farms will often keep nothing but horses for their farm work; for they too will need them for the special carting jobs in any event, and there is not enough heavy work on a small farm to justify the cost of a tractor as well, especially as a tractor must have its own special kinds of implements. Small, irregularly shaped fields, such as are likely to be found on small farms, are better ploughed by horse than by tractor because of the difficulties of turning on the headlands.

The marked advantages of tractors over horses are (i) speed: a tractor can plough three or more furrows at a time where two horses can generally only plough one, and fullest advantage can thus be taken of spells of good weather. Tractors are better, too, for mowing machines and reapers and binders, because these machines can be driven for long periods at high speed; this would tire out the horses, but tractors do not get tired. (ii) Tractors can be used for 'bench work', that is, as stationary engines to drive threshing machines, mills for crushing or grinding corn, and chaff-cutters.

Horse breeding. But in spite of these advantages of tractors, because there are certain jobs where horses are still necessary farmers will continue to breed them. This chapter will describe their methods.

Horse breeding differs from the breeding of any other farm animals for two reasons: (i) farmers do not usually keep on the farm male horses (stallions) of their own as they do bulls, boars, and rams; and (ii) mares are seldom kept solely for breeding, as cattle and sheep are, but are wanted to *work* on the farm as well. The work done by mares, even while they are breeding, helps to pay for their keep, and also keeps the animals healthy and lessens the risk of troubles when their foals are to be born. Stallions, which are kept on special farms, known as stud farms, are not usually worked at all, or at least not during the breeding season.

It is plain that as mares and stallions are kept on different farms and by different owners some arrangement is necessary beforehand to bring them together for breeding. Two systems are in general use in Britain: either the mares are sent to a stud; or a stallion travels round from farm to farm in a district during the mating season. When mares are sent away to the stud they will have to stay there anything from six to twelve weeks to make sure that they are in foal, during which time the farmer will lose their work on his farm and, in addition, will have to pay the stud farm for their food and management. Also the mare may have a foal at foot and there is always the risk that it may be injured while travelling to and from the stud. This method, therefore, is hardly ever followed with farm horses, though it is quite common with valuable race horses or thoroughbreds.

Travelling stallions. Under the other systems, by which the stallion is brought to the mares, a breeder or a horse-breeding society arranges for a stallion to travel round the farms each week during the breeding season. He follows a certain route which takes just a week to complete. Thus during the whole of the breeding season a groom will call with his stallion at any particular farm on the same day every week. The farmer will know the programme beforehand and expect him, and on the day of the week appointed for his farm will have all the breeding mares at the homestead when the stallion visits the farm. Those mares that are on heat will be mated with the stallion; but it is never certain that successful breeding will result from a single mating. Mares are on heat for several days at a time, and it is known that breeding is most likely to follow mating when it takes place at the end of the heat period; but the day when the stallion visits the farm to serve the mares may not coincide with the end of the mare's heat period. The stallion therefore makes weekly visits to the farms to ensure that mating at a successful time shall sooner or later occur.

The stallions continue touring the countryside for about three months (i.e. from April to June inclusive), by which time most of the mares will be in foal. After midsummer most mares cease to come on heat: but in any case farmers do not want foals to be born in the late spring, as they would be from matings later than June.

The pregnancy period. After the mares have been served they work as usual and no change is made for some time in the way in which they are fed. Just after half-way through the pregnancy period a mare will often be found to be quieter in temperament and the presence of a foal may be shown by the enlarging of the belly (though this is by no means always a reliable sign). For the last four or five months of pregnancy the mare should not be worked for long periods without food or water during any day, nor in shafts, which may press on her flanks. Brood mares should be worked in *chains*, and may do such work as harrowing and ploughing. When breeding mares are working hard they sometimes get thin; but a careful horsekeeper will notice the first signs of this and will give extra food to counteract it. The normal ration for such working horses will be about 14 lb. of crushed oats daily, but if a mare is losing condition the addition of 2 or 3 lb. of linseed cake to her feed will quickly check the loss of weight. No change will be needed in the hay ration (14 lb.) and chaff ration (4 lb.).

Foaling-time. As foaling-time approaches it will be noticed that the mare's udder becomes enlarged and filled, but each day when she goes to work it will become partly empty again, refilling when she rests at night. A week or ten days before the foal is expected, the mare should be put each night into a roomy box and well bedded-down with straw. But farm mares are worked during the day right up to the time of foaling, returning at night to their loose boxes where they may be watched as foaling approaches. Of the signs that foaling is shortly to take place, the best indication is the appearance of a waxy substance at the ends of the teats. After this has appeared mares should be watched almost constantly at night time; they will, of course, be seen during the day when they are at work.

Valuable thoroughbred mares are placed in special boxes with peepholes through which they may be watched constantly, day and night, so that help can be given at once if it is needed. The foal is born very quickly, and help, if it is wanted at all, must be given immediately, otherwise complications may arise.

The foal should begin to suck milk from its mother almost as soon as it is born.

Some breeders believe that it is best to tie up a foal's navel with antiseptic tape, others that it is best not to do so.

Management of mare and foal.   Mare and foal will remain indoors for several days, the mare getting plenty of green food, which is naturally laxative. If a foal does not pass the first dung properly (and this is a very common trouble) it may be necessary to give it medicine even when it is only a day or two old.

On the first sunny day after the foal is three days old both mare and foal will be turned out to grass for an hour or two. At first the mare is kept haltered, while the foal moves about quite freely. If the weather remains fine and warm, mare and foal will go out daily, the time spent outside being gradually increased. As soon as the foal is ten days old, and if the weather is fine and warm by day and night, both may stay out on pasture altogether (p. 65).

It should not be forgotten that a mare is usually ready to breed again at about nine days after foaling.

The treatment of the mare and foal after this time will depend upon the amount of horse labour wanted on the farm. Without a doubt, it is best for the foal if the mare does no work during the summer (the suckling period) but stays out on pasture, and provides plenty of milk for the foal. But as this means that the mare will do no work for a period of five months (which includes hay and corn harvest) it is of course expensive. In Norway (though rarely in the United Kingdom) suckling mares are worked whenever they are required, but this definitely reduces the milk yield, and both mares and foals are upset mentally by being separated. In Norway they get over the second difficulty by letting the foal run beside the mare while she is at work.

If, for any reason, a mare has to be used for work while she is still suckling her foal certain precautions must be taken: the mare and foal are taken together into the homestead when the mare is required for work; the foal should be led to a box and carefully shut in so that it cannot jump out; and the box should have no projections or corners that would hurt the foal if it should dash round in the box, trying to find a way out. Two foals will settle down better together than one alone. The foal should not miss its milk feed at mid-day, and so the mare should return at that time. Whenever a foal returns to its dam, the dam should be slightly milked before the foal sucks, because the first milk is poor and might make the foal scour.

Weaning. Some British farmers wean the foals when they are three or four months old; but generally the accepted suckling period is five or even six months. Early weaning may interfere with the growth of the foal, since it does not receive sufficient milk to give it a good start. When foals are weaned at the younger ages it is generally because the mares are urgently wanted for harvest work.

While the mare and foal are still at grass, no additional food is given at first. If because of dry weather the supply of grass should get short, then lucerne, or oats and tares, or clover mixture, may be cut and fed to the grazing mare and foal. As the time for weaning approaches, some crushed oats and bran are given to both animals. This is done (a) to allow the mare to teach the foal to eat from a manger or trough, (b) to prepare the mare for work, and (c) to feed the foal so that it does not receive too great a check when it loses its milk supply at weaning. It is well to have at least two mares and their foals together before weaning because two foals will settle down well together, whereas a single foal is a nuisance after weaning. It may simplify management to sell a single foal or to buy another to keep with it rather than to try to keep it without a companion.

At weaning time some farmers separate the mares from

their foals for increasing periods, until the mares are away for a whole day; the mare being worked, but not for too long, or too heavily at first because she needs a little time to get fit for full work. The foal is usually taken into a homestead at weaning and there kept in for a few days, and the opportunity is taken then to dose it for 'red worm',[1] because it has to be starved for twenty-four hours before it is dosed. After it has recovered from weaning and worming the foal will be turned out to grass again for a considerable time. It should be given some oats and bran in addition to the grass.

Management of growing foals. For the first winter, foals may stay out on grassland. In some districts no shelter is provided, and the concentrated foods are put out daily on the field in a trough. Hay is given only in the hardest weather in the winter. Under another system a shed is provided so that the foals may go in and out at will; and then the concentrated foods and hay will be fed in the shed. Under the third system the foals are kept in open or covered yards (just like bullocks) where they are fed on any or all of the following: hay, straw, concentrates, roots or silage. The yard system developed in the Eastern Counties at a time when bullock fattening in yards was not profitable.

In the spring, when the foals are a year old, the colt foals are castrated, unless they are to be kept as stallions for breeding. Some breeders, however, do not castrate their colts until they are four or five years old; castration has a marked effect on the growth, and this delay allows the geldings to become more massive.

In April or May yearlings are put on grassland. Those that were housed in the winter are turned out on pasture as soon as there is any appreciable growth of grass.

[1] The 'red worm' is a parasite that lives in the digestive tract and when present makes foals very unthrifty.

Throughout their second summer the yearlings normally get nothing beside the grass; but in times of drought additional green stuff may be needed. Sometimes the feet of young horses may get rather broken; on some soils particularly the feet grow too freely; this may upset the development of the legs and cause permanent injury if the feet are not pared to keep them shapely.

For the second winter, the young horses can either remain out at grass or be kept again in yards, according to the way in which they spent their first winter. During this second winter the amount of concentrated foods given daily is reduced to about one-half of the quantity given during the first winter.

In the third spring, the young horses are again at grass, and some breeders will breed from well-grown fillies at this stage. No changes are made in the feeding; the fillies will be served as the mares were, either by travelling stallions or at studs, and after service no distinctions are made in the way in which they are managed till half-way through pregnancy. In the autumn, when the young horses are two and a half years old, it is usual to break them into work, and this is done not only to the geldings but to any fillies that have been served; in fact, experience shows that they will generally be more tractable when in foal. But if the young horses are small and not well developed at two and a half, breaking-in may be postponed until they are three years old.

Breaking-in to work. Though the actual method of breaking-in to work varies from farmer to farmer and depends upon the temperament of the horse, the following general points are important and should be remembered:

(1) A young man usually breaks in horses to walk quickly; an old man breaks them in to walk slowly.

13. Chickens in a battery brooder-house

14. Sheep grazing amongst poultry folding units

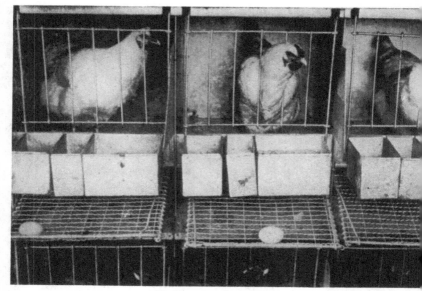

15. Hens in a laying battery

16. Goats and kids grazing the grass verge

(2) Kindness is absolutely essential. Young horses should not be beaten.

(3) A newly broken horse should be brought to full work gradually.

(4) The harness must be a good fit. Badly fitting harness may cause painful sores and so make the horse shy of work for a long time.

(5) Feet should be pared and shod before breaking-in begins.

Some farmers assume that all horses will be difficult to handle and will first harness the horse and leave it for a time in a stall to get used to the feeling of harness. For the next stage the horse is attached to a lounging rein and taken into a meadow (p. 65); one person will stand, holding the end of the lounging rein, and the horse will trot round him in a circle. The lounging rein will train the horse to be managed from the bit and bridle; it also makes it tired and more tractable for the next step, which is to harness the horse to a log of wood, which it will be made to drag about on grass or on a ploughed field: this teaches the horse to pull, and accustoms it to the feel of the weight of the log upon its shoulders. If all goes well, the next step will be either to put the horse into a plough or into a cart. In a plough the young horse is placed beside an old reliable horse, the young horse being hitched nearer to the unploughed land, so that a man can walk beside it and lead it by hand. After a few hours like this the horse will generally work quite well and need no longer be led.

At first it should work for half-days only, so that it gradually gets used to work. After a time, if it is working well, it may then be used in the team of a reaper and binder; though to a horse of nervous temperament the rattle of the binder may turn out to be rather too alarming.

The last stage is to put the horse into a cart. If the horse is expected to be troublesome those parts of the cart within reach of the horse's heels ought to be padded so that if it kicks it will not hurt itself. An empty cart is often used, but this may prove a disadvantage because the rumble of an empty cart is louder than that of a full cart and so may frighten the horse. When a horse will work satisfactorily in both plough and cart it can be described at a sale as 'quiet and a good worker in all gears'. If, on the other hand, it has been used only in a plough then it will be described as 'quiet in chains'.

The main differences found in the process of breaking-in to work depend upon how much trouble the owner is inclined to take. Some even put the horses to work without any preliminaries; others put them into a plough first, and others into a cart first. But the system given in detail above is satisfactory with highly strung horses.

The time taken is roughly as follows: several days lounging and log-pulling; a week or two ploughing for a half-day only so that in several weeks the horse will be ploughing normally; a horse will usually learn to work quite satisfactorily in a cart in a few hours. Some horses break in to work so easily that, as the saying is, they 'go straight to work'; much depends upon the breed and temperament of each horse.

Care of working horses.   Working horses need careful management if they are to work hard and remain healthy. They must be fed regularly; they must receive more corn when working hard than when doing little work. More digestive troubles arise, however, through over-feeding than from any other cause. Horses must have sufficient water, but they must drink at the right time, namely before, and *not* after, eating dry foods.

A well-known saying has it: 'No foot, no horse', and every horse-keeper must pay attention to horses' feet; they must be pared by a blacksmith when necessary, and shoes must be put on whenever the hoof is wearing away, or when the horse is doing much road work. It is bad for horses to stand in wet stables; the liquid softens the hoof and foot troubles may ensue.

With hairy legged breeds the legs must be groomed daily; failure to keep them clean may lead to a disease called 'grease', a form of wet eczema, and incurable. Working horses must be groomed every night and morning, and more time must be given to grooming when they are hot after a day's work. Lack of grooming, or lazy or faulty grooming, leads to skin diseases.

In summer, most working horses live out at grass during the night and for the week-ends. This is healthier and cheaper than if they are kept indoors. The horses are taken to stables for harnessing, and any corn that is to be fed is given to them there.

In the winter there are four alternative ways of housing the working horses. They may be kept—

(a) in stables,
(b) in yards,
(c) in boxes, or
(d) in stables for feeding, and in yards for the nights and week-ends.

Remembering that horses should have exercise, but that if they can have plenty of exercise they may kick each other (though in play), it will be plain that something may be said for and against each system. If plenty of straw is available there is much to be said in favour of (b) and (d). Wherever horses are kept they need litter for nights; and with stables the bed must be cleaned up daily.

6-2

No mention has been made in this chapter of the lighter horses, but the same general principles apply as have been described for heavy horses, except that light horses are not broken in to work in the plough, but to riding or into a cart or trap. Light horses are inclined to be spirited, and may rear up and kick a good deal. Special harness is often used to control them and to prevent this. Breaking-in light horses to jump and to race involves quite a different method, but hardly comes within the scope of a book on farm animals.

CHAPTER VIII. POULTRY

The old system. Nothing on the farm perhaps has changed so much in the last fifty years as the methods of poultry keeping. The old-fashioned way of keeping hens was simply to have them run about the place and fend for themselves in the stackyard. Each hen probably laid about fifty eggs in a year, and lived on till she died a natural death. Their nests were where they chose to make them—in odd corners of the stackyard—and often the first evidence of their laying was seen when a hen appeared importantly leading half a dozen or so chickens. Haphazard and uncertain as the method was, still the farmer's wife got some eggs from her poultry, and now and then a bird was killed for the table, if a hen was thought to be broody, or a cockerel was one too many in the flock.

In this old system (if it could be called a system) no control at all was exercised over the production of eggs and poultry. It has been replaced by a number of different systems in all of which the aim is to control accurately and profitably the breeding and laying and fattening for table. In the extremest modern systems, by contrast, the poultry flock is not only run like a machine, it may be in fact partly run by machines: eggs are hatched in incubators, the chicks are reared in wire cages (p. 80), moved to the range (or run) for a growing interval, and moved on from there to other wire cages where they live and lay their eggs (p. 81). Sometimes not only have food and water been taken to the birds by a moving-belt system, but eggs collected and cages cleaned out and the droppings removed, all by moving belts or conveyors. This makes it clear enough that poultry keeping has seen great changes, though perhaps not all the changes tried have been justified.

Breeding. Under the old system cocks were kept constantly with the hens, and in the spring, whenever a hen was broody, about thirteen eggs were selected and placed under her. This hen was put in a box and kept more or less in the dark. She was allowed off her nest once a day for from ten to thirty minutes for food and exercise. Sometimes the hen forsook her eggs; if she left them for more than an hour or two they got cold, and when once cold the eggs were spoilt. If all went well, however, after three weeks the chickens would hatch out, and the hen and her chicks were then moved to a hen coop in some such convenient place as an orchard, the stackyard, or even on pastures away from the biggest stock, such as horses and cows. After a fortnight the hen was let out of the coop to wander and forage with her chicks. She would remain with them for about two months, after which hen and chicks were separated, and all the chicks amalgamated into groups according to their sex.

According to more modern practice, much developed in recent years, the cockerels are run on pastures along with the hens, but not with the pullets, because if pullets' eggs are used for hatching the resulting stock seems to lack vigour.

This outdoor system is good and produces healthy, vigorous stock if care is taken that the hens be kept on *fresh* grassland, that is to say on grassland that has not been used for poultry for two years.

Sometimes even to-day the fertile eggs from these pastures are put under broody hens to be hatched, but much more commonly incubators are used for hatching.

Hatching. An incubator is simply a container for eggs with some arrangement by which the eggs in it can be kept at an even and correct degree of warmth. The smallest incubators have a capacity for 50 eggs; the biggest may take as many as 20,000 at a time. They may be heated by oil, or gas, or electricity. It is most important when building an incubator to make sure that it stands perfectly level, because even a small slope may cause a faulty flow of air and lead to a hot spot or a cold spot in the incubator, either of which faults will spoil the eggs that lie in that spot.

Eggs must be turned twice every twenty-four hours; in the smaller incubators by hand, but in big incubators automatically. As evenness of temperature is one essential for successful hatching, so a proper degree of humidity (or dampness) is another, and it may be necessary, by sprinkling water for example, or by housing the incubator suitably, to take precautions to ensure this.

After a few days in the incubator the eggs are 'candled', which means that a powerful electric light is made to shine through each egg in turn so that those which are fertile, and have a chick starting to grow inside them, may be returned to the incubator, and those that are clear discarded. If eggs are candled after no longer than seven days in the incubator those that are clear need not be wasted but can be used for food. But it is necessary, under this system, to candle the remaining eggs after another seven days, that is, at their fourteenth day in the incubator. As candling is a slow, egg-by-egg process, some breeders find that it is more profitable (even though it involves wasting a certain number of infertile eggs) to candle them once only, at ten days. Then the fertile eggs are returned to the incubator, and all the rest discarded.

After a total of twenty-one days the chicks should hatch out, and most incubators are now designed so that the newly hatched chicks can leave the shell behind and move to a special drying platform. The first chicks to hatch are always the strongest, but it is quite satisfactory to rear all chicks that hatch within twenty-four hours of the first appearing. Most hatcheries do not trouble to rear any chicks that are later than this. While they are in the incubator they need no food, for every chick brings with it out of the egg a supply of food inside it (it comes from the yolk), enough to last several hours.

Rearing. When it is time for the chicks to leave the incubator they are moved into a 'foster mother', a device that does what in fact the hen herself would do if the chicks were being naturally hatched and reared, it keeps the chicks snug and warm. The principle of all 'foster mothers' or brooders is the same—simply a heating device warming the space where the chicks are kept. It may be a simple oil stove standing on the floor of the deep tray where the chickens are kept, with a special cowl to direct the heat to the floor about them; or the same arrangement, with an anthracite stove instead of the lamp. If more room than this is needed, as it often is in the big hatcheries with their thousands of chicks, then the 'foster mother' (p. **80**) may be a whole room, specially constructed, and kept at a constant temperature by a hot-water system. The temperature in a brooder has to be regulated according to the age of the chicks; the younger the chicks, the higher the temperature. Another kind of brooder consists of a series of deep trays arranged above one another in stands in the warmed room. The youngest chicks are in the trays at the top and the oldest at the bottom, the degree of warmth being naturally highest at the top and lowest at the bottom. As chicks are hatched they are put into trays high up, and are moved down one level each week so that in six or eight weeks in this way they reach the floor, and are ready to be moved into another building.

At about six weeks of age the pullets and cockerels are separated, and the pullets are put on pastures in small huts called Sussex Arks, which are there for shelter only. No food is given in the huts and no nest boxes are necessary, because the pullets are too young to begin laying, but the outdoor method is a healthy way of rearing them. The field, however, should not be used for rearing more often than a spell of six months every two or three years. These pullets will remain in their Sussex Arks for two months or so, the exact time depending on the breed: small breeds take the least time. The separated cockerels are put into other Sussex Arks, but are reared on fields away from those used for pullets. Here they will remain until they are needed for breeding or for fattening. Fattening for table takes anything up to two months and will be described later.

Laying birds. Pullets begin to lay when they are about six months old, though the actual time may be a little shorter or longer according to the breed and the system of feeding. Small breeds (such as Leghorns) begin at five months, and heavier breeds (such as Light Sussex) at six or seven months; Buff Orpingtons do not start laying until they are seven or eight months old. It is well known that feeding with high-protein foods leads to early laying, and that carbohydrate foods delay the laying age.

About a month before laying is expected pullets should be moved into their 'houses' or laying quarters; if they are not moved until laying has commenced egg production may be upset.

Housing of pullets and laying hens. Too often in the old days any spare shed in the stackyard, or even a rough shelter in odd corners among other buildings nearby, would be fitted with rails and nesting boxes and the pullets and hens put there to lay. Such places were probably badly ventilated and seldom cleaned out, and parasites were very abundant. The pullets moved into these places as they reached the laying age, caught all the parasites and diseases to be found there. So haphazard a way of treating laying hens could never be successful. Space, ventilation, and cleanliness are essential to keep the birds healthy, and, naturally, birds that are not healthy can never be the best of egg producers.

Big laying houses. During recent years really large laying houses have been developed. These are fitted with roosting racks or benches, nesting boxes, and troughs for meal, grit, and water; the corn that is fed is usually sprinkled on the floor, on which chaff has also been spread. Large windows are provided to give as much *light* as possible, and it has also been found that if in winter time the houses are lit by artificial light to prolong feeding time the birds will lay more eggs. Various lights, including electric light, have been used for this purpose, controlled by some clockwork device which turns them on and off at the times decided.

If these large laying houses have no runs they must have plenty of floor space; if, however, grass runs are provided less floor space is needed, and corn can be fed outside on the pasture. When huts are built with runs the best plan is to arrange matters so that two alternative grass pens may be used; the grass in one pen may thus be rested while the other is grazed.

Portable laying houses. These are small, holding only from sixty to a hundred birds, and are built on wheels so that they may be easily shifted. They should be moved to fresh pastures every three months, or at least every six months, and in the autumn they are taken to stubbles, where the hens can find grain, weed-seeds, and insects. As the birds are free to roam the fields during daylight this system of portable houses without pens cannot be used in districts where there are many foxes.

Folding units. These are still smaller and lighter structures; they provide complete living quarters for the one or two dozen birds they house, since they have, beside the house with its roosting rails and nest boxes, a limited run in which the birds can graze (p. **80**). But the area of grass in the run is so small that the whole unit must be moved over to fresh grass once or even twice daily. The units are, however, so lightly made that one person can move them the necessary small distances. Each hut is moved across the field its own width each time, so that the whole field is grazed and manured in an orderly manner. Portable units like this have been largely used in districts where there is risk of foxes, and on light soils where free drainage makes it possible for the birds to be on the grass in winter as well as in summer and where the manuring would improve the productivity of the soil.

Battery laying houses. These are a series of cages arranged in rows in tiers, from two to five rows above one another. Each laying bird is in a pen to itself (p. **81**), and it remains in this pen till it is culled and killed for the table. The laying pens are small and give the bird barely more than room to turn round. Food, grit and water are constantly before the birds, and it is arranged that as eggs are laid they fall on to a sloping platform and roll forward out of the hen's reach. If a pen is so built that the hens can reach the eggs they may eat them. Birds kept under these conditions get no exercise and may not be healthy; it is generally recognized, therefore, that battery-kept birds should never be used for breeding.

Breeding pens. A breeding pen consists of a small hut and a grass run. Ten or twelve carefully selected hens are kept in such a pen and are mated with a valuable cock. The hut may be fitted with special trap-nests. When any hen enters one of these nests to lay she is trapped there and can only be released by hand; each hen is thus credited with her own eggs, and the parents of every egg are known. The cock will remain with the hens as long as eggs for hatching are needed. Great care is taken with these special eggs, which are often hatched not in incubators but under broody hens and then kept in coops on grassland. If an incubator is used then each egg is marked with the identification number of the hen that laid it, and, as hatching time approaches the special egg is put into a small numbered wire cage in the incubator. As soon as the chick hatches a numbered metal band is fixed to one of its wings, and when, later on, this number is removed, a numbered ring is fixed to the leg. By this method the parentage of each chick is known, and the breeder can be assured that the young bird comes from healthy and pure-bred stock.

Rearing will proceed as with the other chicks, unless the breeder for some reason decides to give them separate and closer care. The main purpose of these breeding pens is the production of pure-strain and healthy cockerels for breeding, but all healthy pullets will be saved also, and only those that are found to be faulty will be fattened and killed.

Fattening. The youngest birds to be killed are generally the male chicks of such breeds as are of little use for meat production, because if these chicks were fed and allowed to grow they would never make satisfactory table birds. They are killed off as an economy in feeding as soon as they can be distinguished from the pullets.[1] In some countries these young birds are in demand as delicacies; in others they are wasted because no one will eat them. Such small birds are usually skinned, not plucked.

Petit poussin, fattened birds of six to eight weeks old, are in peace time a special delicacy and when the game birds go off the market in February these command such good prices that some poultrymen make a special business of rearing them. Petit poussin are fattened immediately the cocks can be distinguished with certainty from the pullets, that is at about six weeks. Fattening is done indoors, a number of birds together.

When birds are sold fat at sixteen weeks old they will usually have spent a month or two under range conditions. Birds for this grade are first kept in a building or in a hut with a confined run for a matter of two months, spending their last three weeks under more intensive fattening, during which time they may be in pens and allowed no exercise at all. Sometimes half a dozen birds or more (the number depending upon their size and the size of the pen) are packed together into a pen. In some districts, especially in Sussex, they are 'crammed' to get them fat. Cramming is done by a special forcing machine which pumps more food into their crops than they would otherwise pick up. It is an ancient system, believed to have been first practised in this country in Sussex by the Romans.

For the Christmas trade, when big birds are needed, chicks twenty weeks old or more are fattened. These birds get a course of intensive fattening in close pens, just as described for the sixteen weeks' class, except that they undergo a *two months'* fattening period.

Older birds than this are also killed and sold for the

---

[1] With some crosses sexes can be seen at hatching because of variations in colour due to sex, but some experts can determine the sex of a chicken by looking at its cloacal opening.

table. These are merely those birds which are taken from the flock of laying hens because they have stopped laying. Thus they are often not fattened at all. Many table fowls in the market are of this class, for at least half of the number of pullets that enter the laying pens in the autumn are killed before the summer comes, and of those hens that are kept for the second laying winter probably all will be killed off and sold before the next laying season. Culling in both of these classes of laying birds takes place continuously throughout the season.

This short sketch of the more up-to-date methods of breeding, hatching, rearing and fattening will show how important an orderly system has become, and how little is now left to chance. Specialization as the result of knowledge has brought much greater efficiency; the birds fatten more readily, or lay more eggs, than they ever did in the old haphazard barnyard flocks. With good management a whole flock will produce on the average 160– 200 eggs per bird per year, while occasional birds will lay over 300 eggs in a year. This is very different from the 50 or so laid by the farmyard hens.

CHAPTER IX. GOATS

GOATS produce good milk; they are fairly easy to feed and to look after, and as they can be kept in pairs or even singly they are particularly suitable to be kept by any country householder. It is not surprising, therefore, that goats become popular in war time.

Housing in winter. In the latter part of winter most goats are dry (i.e. they give no milk), and at that time are kept under cover, because although cold does not hurt them they cannot stand rain, or wet conditions. No elaborate building is necessary; the roof should be good and the floor dry, there should be a special manger with a hole cut in it to take a bucket for water and a rack for hay. Some people do not use a hayrack but put the hay and leaves into nets; the goats can then pull their food from the nets, without wasting it as they would if it lay on the floor to be trodden. The goats are usually kept tied up in their house by a rope round the neck; they will need exercise and should be led out for this daily or, in fine weather, they may be tethered on grassland.

Rearing kids. Goats have their kids in the spring, from March till June or July, as sheep do. They also have a definite breeding period (September to February) and carry their young, as sheep do, for five months. There is usually little or no difficulty when the kids are born, and no special precautions are needed. Modern breeds have been so much improved that they often have two and sometimes three kids at a time.

Only the female kids are likely to be reared, so unless a male kid is wanted for breeding it will be either killed off at birth or fed until it is six to eight weeks old, and then slaughtered for meat. Such kids as are kept for breeding purposes may sometimes be suckled by the nanny for a few weeks, but most people believe that this seriously affects milk production (which is, of course, the main purpose for which goats are kept) and the commonest practice is therefore to take the kids away at birth and rear them by bottle feeding. Leaving the kids to be suckled by the nanny certainly causes uneven development of the udder.

To rear kids on the bottle successfully careful attention must be paid to certain details. For the first ten days milk should be given three times a day, half a pint at each feed. Goat's milk, of course, needs no dilution, but if cow's milk is given it should be mixed with water in the proportion of one part of water to four parts of milk. For the second ten days the feeding is done twice a day, with one pint of the milk or the milk mixture at each feed. For the third period of ten days, times and quantity remain the same, but the cow's milk is given without any water mixed with it. After the third period of ten days the amount of food given daily is gradually reduced, until, by the time the kids are six weeks old they are getting no milk at all. Sometimes as the milk is reduced porridge is given for a while to make sure that the kids get enough to eat. At three or four weeks old they will have started to eat some solid foods such as dandelions, clover, hogweed, and tree branches and their leaves. At the same time some concentrates should be given; bran is very good, or bran and flaked maize mixed, one part of flaked maize to eight parts of bran.

Rearing at pasture. At weaning time the female kids may be run together at pasture for a while, but it will not be long before they begin to show their natural tendency to eat bark and they may damage any trees on their pasture. There is always the likelihood, too, that they will find or make gaps in the hedges and so stray away. So at three or four months old it becomes necessary to tether them, and from that time they will be tethered whenever they are out of doors. The best tether is a chain, but if rope is used a short length of chain should be used near the goat's head, for a goat will inevitably gnaw through any rope that is used for the head or neck. It is a common sight to see goats tethered on the grass at the hedgerows and grass verges at the roadsides (p. 81), and this is a very satisfactory way of making use of that kind of grassland.

If it has not been done before, the young billies and nannies should be separated in the autumn, which is the natural breeding season, or else they may breed. It is better for nannies not to breed in their first winter, when they are under one year old; their own growth will be retarded, they will only produce one small kid and a little milk, and thus nothing will be gained.

When the nannies are yearlings they should be tethered out whenever the weather is suitable. No concentrated foods are now necessary, but if they are indoors then prunings from fruit trees, and tree leaves if available, may be given in addition to the hay.

Breeding. As autumn approaches the nannies should be watched for signs that they are ready to breed. If they are constantly bleating and wagging their tails they are ready for the billy, and if the first mating is not successful they will breed again three weeks later. The normal breeding season is from September to February, and it is noticed that if a billy goat is kept near, the goats will come into season earlier. Kids are born five months after mating; thus in February the first kids will arrive. No special food need be given to the goats just before kidding, and it is rare for them to need any help when

their kids are being born. Treatment of the newly born
kids has already been described.

The milking nannies. The milking is carried out
in various ways. The practice on the Continent is to milk
the goats from behind, but in the British Isles from the
side. The milking is done into low buckets or even pans.
Sometimes the goats are placed on a raised platform to
make milking easier.

The amount of milk given by goats is astonishingly
large, as much as one gallon a day being quite commonly
obtained. Two gallons is the record yield. The quality
of the milk varies considerably, the amount of butter fat
may be anything between 4 per cent and 9 per cent.
Goats should be milked twice daily, and any giving more
than three pints at a milking, three times daily.

Some people will not drink goat's milk simply because
it is goat's milk; others, with a better reason, say they
dislike it because it carries a noticeable 'goaty' flavour.
It seems to be established that this taint is in some way
connected with the presence of billies near the milking
nannies, and experienced goat keepers declare that if no
billy lives with or near the nannies no such flavour is
found in the milk.

Goat's milk is free from tuberculosis, and for this good
reason is recommended for children. Some households,
indeed, keep goats mainly for this reason. Cheese and
butter can be made from the milk, but are not popular
in the British Isles where almost all of the small total
quantity of goat's milk produced is used as liquid milk.
Goat's milk butter is very white by comparison with
ordinary butter.

The further important points of management of the
milking herd may be briefly noted:

(1) Goats are liable to be infested with lice, and this

should be watched for. It can be cured by treating them with any wash sold for the purpose and containing derris powder.

(2) If goats are kept on soft ground their hoofs will get long, and it may be necessary to clip or pare the hoofs to prevent lameness.

(3) Food consists of hay in any quantity, bran and cereals; and in the summer time goats can be used as scavengers on any waste land, but tethering is essential.

Keeping goats is not costly or difficult or laborious work; the goat converts foods that would otherwise be wasted into milk for human food—a very good reason why the goat should be as popular as it is at the present time.

Printed in the United States
By Bookmasters